# Design Art of VILLA
## IV
### 豪宅设计典范 IV

深圳市艺力文化发展有限公司 编

华南理工大学出版社
SOUTH CHINA UNIVERSITY OF TECHNOLOGY PRESS

·广州·

图书在版编目（CIP）数据

豪宅设计典范 = Design art of villa.4：英汉对照 / 深圳市艺力文化发展有限公司编 . — 广州：华南理工大学出版社，2015.6
ISBN 978-7-5623-4612-8

Ⅰ．①豪… Ⅱ．①深… Ⅲ．①别墅－室内装饰设计－世界－图集 Ⅳ．① TU241-64

中国版本图书馆 CIP 数据核字（2015）第 081175 号

**豪宅设计典范 Ⅳ Design Art of Villa Ⅳ**
深圳市艺力文化发展有限公司　编

出 版 人：韩中伟
出版发行：华南理工大学出版社
　　　　　（广州五山华南理工大学 17 号楼，邮编 510640）
　　　　　http://www.scutpress.com.cn　E-mail: scutc13@scut.edu.cn
　　　　　营销部电话：020-87113487 87111048（传真）
策划编辑：赖淑华
责任编辑：陈　昊　黄丽谊
印 刷 者：深圳市汇亿丰印刷科技有限公司
开 　 本：595mm×1020mm　1/16　印张：21
成品尺寸：248mm×290mm
版 　 次：2015 年 6 月第 1 版　2015 年 6 月第 1 次印刷
定 　 价：360.00 元

版权所有　盗版必究　　印装差错　负责调换

# PREFACE
序言

The origin of the Villa dates back to Roman times when "Villa" referred to an upper class country house. From antiquity through the modern era, Villa implies not just living in the country, but being implicitly removed from the city, on the outskirts, becoming one with nature and the land. Certainly, Villas throughout history have maintained this quality, they spread out and they embrace the land.

Spreading out as they do enables the resident great latitude to accomplish much in the way of features, which one cannot obtain in an urban residence, "Domus". Villas commonly contain areas that accommodate many sleeping quarters, varied living spaces both indoor and outdoor, and promote leisure activities with amenities such as tennis courts and swimming pools.

In the recent past, Le Corbusier and Mies Van der Rohe, two of the 20th century's architectural masters, clearly understood these ideals when they titled their early breakthrough modern homes "Villa Savoye" and "Villa Tugenhut". And though, not specifically titled as such, the compositions of Frank Lloyd Wright's "Falling Water" and Walter Gropius's home in Lincoln, Massachusetts, clearly embody the characteristics of the Villa.

It is in this context that the modern homes, or Villas, represented in this book, carry on the traditions. Removed from the city embracing the landscape, spreading their wings, creating vistas to the sea, and openings beyond the common scale of doors and windows. Some Villas anchor to the ground as if to say: "I am here to stay," "I am one with you". Others express a sense of soaring and hovering, they float above the landscape respecting its independence. From the interiors to the exteriors, the modern Villas in this book are truly a work of art.

Stuart Narofsky, AIA, LEED AP

别墅的起源可以追溯到古罗马时期，当时"别墅"是指上层阶级拥有的乡村住宅。进入现代后，别墅一词不仅指乡间住宅，更多了一层远离城市喧嚣、向郊区转移，融入大地与大自然的含义。纵观别墅历史，它基于大地、拥抱大地的特质始终未变。

别墅规模较大，有许多鲜明的特色，居住者可以自在地居住其中，享受即便是都市住宅"Domus"也无法给予的生活乐趣。别墅常含多间卧室和多样的室内外空间，如网球场和游泳池等，来支持各种休闲娱乐活动。

两位二十世纪的建筑大师勒·柯布西耶和密斯·凡·德罗近年来深谙这些理念，他们在房屋建造上实现了突破，打造出"萨瓦伊别墅"和"图根哈特别墅"。而弗兰克·劳埃德·赖特的"流水别墅"，以及瓦尔特·格罗皮乌斯的马萨诸塞州林肯住宅虽较少提及，却同样鲜明地体现了这些特质。

这本书中的现代住宅和别墅继承了这些传统。它们远离城市的喧嚣，自由延伸，与景观融为一体；同样的门窗规模，却使人可以看得更远，感受大海的浩瀚。有的别墅扎根于土地，仿佛在说，"我在此守候着"，"我陪伴着你"；有的别墅则独占一方，高高耸立，似乎于周围景观上空翱翔盘旋。书中的别墅从室内到室外，都堪称艺术品。

斯图亚特·纳洛富斯基
美国建筑师协会会员及绿色建筑认证专家

# CONTENTS
目录

1232 Sunset Plaza    002
日落广场 1232 号

9133 Oriole Way    010
黄鹂路 9133 号别墅

Sands Point Residence    022
桑茨波恩特别墅

Harker Street, Plettenberg Bay    042
普利登堡湾哈克街豪宅

LV House    050
LV 屋

Head Road 1818    110
首脑路 1818 号别墅

438 N. Faring Rd    118
法林路 438 号别墅

Winelands 190    126
190 号酒庄

Expressing Views    136
观景别墅

Laurel Way    142
月桂街别墅

San Vicente Residence    062
圣文森特别墅

Sunset Strip Residence    074
日落大道豪宅

Butterfly House    084
蝴蝶屋

Cliff House    092
崖居

Cove 3    102
海湾三号

River House    150
河畔别墅

Fieldview    158
Fieldview 别墅

Wind House    164
风之墅

Cadence Residence    174
卡当斯别墅

Aloe Ridge House    182
Aloe Ridge 别墅

Float House 192
悬浮小屋

Butternut 200
白托纳特宅邸

Oxford 49 210
牛津49号别墅

Wallace Ridge 218
华莱士山脊别墅

Clovelly House 226
克劳夫利别墅

Kona Residence 238
科纳别墅

House Boz 244
博兹别墅

Residence Amsterdam 252
阿姆斯特丹住宅

Seacliff Residence 258
锡克里夫别墅

Albizia House 266
合欢屋

Daniel's Lane Overview 276
长道盛景别墅

Balcony House 284
阳台屋

Private Residence 294
隐蔽的豪宅

Russian Hill 300
俄罗斯山别墅

Casa Sorteo Tec 191 310
191号现代别墅

Villa Escarpa 316
艾斯卡帕别墅

Oceanique Villas 322
海洋别墅

CONTRIBUTORS 326
设计师名录

国际视野新站点 | 案例丰富新颖 | 访谈顶尖设计 | 挖掘新锐设计师 | 国际设计界缩影

## ■ ACS 创意·空间　　　ABOUT

十年专注于建筑、室内、景观和平面设计，业务横跨图书出版、发行、文化传媒、品牌运营及艺术品市场等多个经营领域，Artpower 自版发行 600 多本图书，收揽全球顶尖设计公司和设计师近 40 000 套优秀原创作品（不断更新ING）。

ACS 整合 Artpower 线上资源，推荐前沿创意理念、概念性设计思维；发布创意赛事活动；组织设计大师访谈；展示新锐设计师作品，推介设计项目；提供私人定制出版和众筹出版等服务。与国际设计团队对接，在全球范围内打造专业设计师展示和交流平台。

## ■ 我们能做什么？　　　HOW

注册成为网站会员，做个人网页，独享会员特权；上传个人作品，展示设计理念，交流成长，互通合作契机。

登录浏览，尽享 40000+ 海内外设计大师作品；建筑、室内、景观、平面、产品、环境设计等分门别类，轻松导航，应有尽有。

## ■ 挖掘新锐设计师　　　DESIGNERS

· ACS 线上展厅
我的 ACS 我做主！设计师可以尽情发表自己的作品，让世界各地的设计师共同关注你的成长！

· 设计师发布会
如果你还在为身为"新人"的标签所困扰，ACS 展示平台只有"新锐设计师"。把曾经因各种原因被否掉的方案重新发表出来，也许你就是那个我们要找的设计师！

· ACS 把设计师的项目推送给全世界，设计无国界，一起关注和交流！

### DESIGN FOR DESIGN

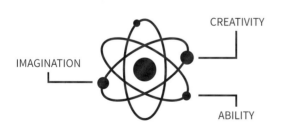

## ■ 私人定制出版　　　PRIVATE

ACS 创意空间联营平台为您提供私人定制出版服务。

## ■ 线下俱乐部　　　ACTIVE

ACS 创意空间俱乐部，不定期邀请国内外顶尖设计师，举办各种创意设计讲座、创意沙龙等，分享天马行空的有趣创意，是思维碰撞、灵感横溢的场所，是趣味相投、惺惺相惜的交友平台，也是企业品牌的展示空间。有机会成为线下俱乐部盟主！

---

设计 · 杂志 · 中英文 · 双月刊
Artpower Creative Space（ACS）创意空间（245mm×325mm · 168 页 · 68CNY）

《ACS 创意空间》杂志是 Artpower 倾力打造的高端空间设计专业期刊。中英双语，全球同步发行；单期发行量逾万册，更有黎巴嫩等国家的专售版；装饰行业至佳交流平台，传播设计新锐资讯；高端空间设计专业期刊，发布国际最优秀室内设计师和建筑设计师的最新作品。

深圳市艺力文化发展有限公司
艺力国际出版有限公司（香港）
深圳市艺力文化发展有限公司北京分公司
深圳市艺力文化发展有限公司厦门分公司

出版合作 / 广告合作：rainly@artpower.com.cn（王小姐）
作品投稿：artpower@artpower.com.cn（莫小姐）

艺力 ACS 创意空间
扫描即可关注！

全球优质设计作品，
尽在 ACS 创意空间

WWW.ACS.CN

最新锐的创意灵感　　最前沿的设计概念

《ACS 创意空间》—— 国际青年设计师协会官方指定合作媒体

关注有惊喜！

扫描二维码，开启电子阅读体验，海量优秀作品随心看！

# 1232 Sunset Plaza

日落广场 1232 号

**Architects**
Belzberg Architects

**Contributor**
The Agency

**Location**
Sunset Strip, Los Angeles

**Photography**
Jim Bartsch

Only on a rare occasion, we find a drive-on estate in the Sunset Strip of this scale and capacity, located just seconds from the famed boulevard's array of world-class restaurants, shopping and nightlife.

To journey beyond the grand gates and soaring pepper tree hedge of this Hagy Belzberg-designed compound estate is to enter a secluded, resort-like sanctuary where three distinct structures — a Main Residence, Wellness Center and Guest House — summon you forward, beckoning you to journey through their space so they can unveil one surprise after another.

Architecturally inspiring, in both appearance and function, 1232 Sunset Plaza radiates a warm, California modern allure draped in sophistication and delight, while

conveying an unmistakable sense of strength, volume and boldness. Exterior and interior transitional spaces are rich in fluidity so that the capacity to enjoy and entertain is always sensible and effortless.

Sweeping panoramic views of the city, afforded by an endless number of vantage points throughout the estate, are nothing short of breathtaking. Equally stunning are the views of the property itself from within, as architectural and landscape elements serve as artistic expressions and repeatedly delight as one explores the property.

  该房屋的规模是日落大道上其他房屋无法相比的。而且，建筑所处地理位置优越，开车能直接抵达，几步开外就是著名的高档餐厅、商店和夜市。

  这栋混合式建筑由建筑师哈基·贝尔兹伯格（Hagy Belzberg）打造而成。高耸的胡椒树篱笆将建筑包围，而隐蔽的度假村式空间则隐藏在大门里面。场地内包含三大主要结构：主楼、健身中心和客房楼。探索建筑之旅引人入胜，将给你无限惊喜。

  不论是建筑的外观，还是功能，都令人赞叹。房屋不仅温馨高档，而且具有一种加利福利亚现代风格魅力；不仅能使人心情愉快，还能给人一种力量感、空间感和无畏感。此外，室内与室外的过渡空间也十分流畅，使得享受和娱乐更加轻松便利。

  住宅的观景点数不胜数，而且各处都能俯瞰城市迷人的风光。住宅本身也是一道亮丽的风景，它蕴含的建筑元素和景观元素都是艺术的表现。探索者能从中不断获得惊喜。

# 9133 Oriole Way

**黄鹂路 9133 号别墅**

**Contributor**
The Agency

**Location**
Sunset Strip, Los Angeles

**Site Area**
1,849 m²

**Photography**
Simon Berlyn

With sweeping, panoramic views of the entire Los Angeles basin, this stunning contemporary architecture is perched on celebrity-studded Oriole Way, in the highly desirable "Bird Streets" above the Sunset Strip.

The brand new estate was meticulously designed to deliver clean lines with wide open spaces, walls of glass and Fleetwood pocket doors throughout that seamlessly fuse the interior and exterior, offering the ultimate California lifestyle. The materials, details and natural light are exquisite.

Owner Sean Sassounian, in close collaboration with top design firm In-Ex, focused intently on the well-curated interior, sparing no expense. The impressive list of European manufacturers includes furniture by Acerbis, Arco, Classicon,

Glas Italia, Matteo Grassi and Walter Knoll; custom lighting by Foscarni; closets by Molteni; outdoor furniture by Paola Lenti, Kettal and Roda; and laundry room, kitchen and pantry by DADA. In addition, the art in the house is specially curated by the Michael Kohn Gallery — with notable and emerging artists, many from California.

This is truly a designer home. As such, all furniture is included in the sale price. Main level offers a spacious grey and white lacquer DADA kitchen that flows to attached sitting/media room, facing out to the pool and lush hills. A massive marble island in the kitchen complements the integrated Miele appliances, and a hidden door leads to an entire catering kitchen behind the main kitchen. Formal living room with high-ceilings and spectacular views boasts a full bar and flows to a gorgeous master office with floor-to-ceiling windows. A sophisticated library/media room provides views over West Hollywood and the Sunset Strip. Tasteful built-in cabinetry throughout. Extremely private, even with the glass and indoor/outdoor flow.

Resort-style backyard offers an infinity edge pool & spa, outside patios with 270° birds-eye views, complete outdoor kitchen, full bath and a spectacular dining and entertaining area. Outdoor kitchen has all Viking appliances and stainless steel Viking cabinets.

Upper level, enjoy 4 bedrooms with wrap-around windows and automated blackout blinds — you feel like you're on an island in the hills floating above the city. Hidden TVs drop from the ceiling. Views everywhere, even the master walk-in closet has a large window and amazing view. All bedrooms are en-suite and feature walk-in closets.

/ 013

Master suite boasts a huge custom master bath with views overlooking the back pool and canyon, dual rain-heads, deep soaking tub & spa, a mesmerizing hall of mirrors effect. Master bedroom faces west for incredible sunsets and afternoon light. Enjoy a private balcony overlooking pool and an attached office and wet bar.

Lower level, find a full gym with shower, separate massage room, powder room and full bath; wine cellar with its own bar; screening room with custom-installed marble bar-counter; en-suite guest room; and a spacious garage with room for 8 vehicles, plus a carport for 2.

The grand entrance of the home features lush grounds, a fountain, custom blended grey terrazzo floors, and a massive American oak door behind double gates. This is a truly chic, one-of-a-kind home on one of the best streets in Los Angeles.

这栋当代风格别墅位于相当抢手的"鸟街"之———黄鹂路 (Oriole Way),此处明星云集,离日落大道也不远,并能饱览整个洛杉矶盆地风景。

该全新别墅的设计一丝不苟,清晰的线条、开敞的空间、玻璃幕墙,配上弗利特伍德(Fleetwood)折叠门,室内室外空间连为一体,没有丝毫瑕疵,鲜明地展现了加州生活方式。同时,别墅还采用了精心挑选的材料,细节处理细致到位,自然采光充足。

房主肖恩·萨索尼安(Sean Sassounian)与顶尖设计公司In-Ex密切合作,不惜成本,精心打造了别墅的室内空间。别墅选用了许多制造商的产品,其中令人印象深刻的欧洲制造商包括Acerbis、Arco、Classicon、Glas Italia、Matteo Grassi和Walter Knoll。定制的灯具是由Foscarni提供,衣柜来自Molteni,Paola Lenti、Kettal和Roda提供了户外家具,DADA提供了洗衣间、厨房和餐具室设施。而且,室内的艺术品是由迈克尔·科恩画廊(Michael Kohn Gallery)提供,都是由著名艺术家和新艺术家创作,大部分是当地的艺术家。

别墅设计精湛,选用的家具价格也合理。主层的厨房十分宽敞,并以灰色和白色DADA厨房设施装饰,独立大理石灶台与Miele厨具相呼应,穿过隐藏式门便是

主厨房里边的烹饪厨房。厨房与客厅／媒体室连在一起，从中可以眺望水池和葱郁的小山。正厅的天花板很高，把壮丽的风景引入室内，客厅内设有完整的吧台，与之相连的是被玻璃幕墙包围的主办公室，非常华丽。高档的书房（媒体间）能直接鸟瞰西好莱坞和日落大道。步入式橱柜贯穿了整个空间，雅致美观。虽然空间设有许多玻璃门窗，但是空间仍然保证了充分的私密性。

度假村式的后院中设有无边泳池和水疗池。此处的户外露台能270°环视周围的风景，还有设施完善的室外厨房、浴室以及风景独好的用餐、娱乐区域。另外，户外厨房中设有 Viking 厨房电器和 Viking 不锈钢橱柜。

二楼共有 4 间卧室，每个空间设有环绕式的窗户和自动百叶窗，给人一种漂浮于城市之上、独立于山间的感觉。隐藏式电视悬在天花板下。从空间的各个角落都可欣赏美景，甚至是主步入式衣柜都设有大窗户，窗外便是迷人的风景。每间卧室都是套房，并配有步入式橱柜。

主卧室配有宽敞的浴室，在沐浴其中的同时还能俯瞰后院的水池和峡谷。浴室设施包括双喷头、深浴缸、温泉和具有镜面效果的过道。主卧室面朝西面，能欣赏迷人的日落风景和午后阳光。独立阳台下方是水池，隔壁则是办公室和小酒吧。

地下一层中是设有淋浴设施的健身馆、独立按摩室、盥洗室和卫生间；其中的酒窖配有酒吧；放映室中设有定制的大理石吧台；还有会客室，以及 8 车位超大车库和 2 车位停车场。

别墅前面是葱郁的绿地、喷泉和专门打造的水磨石露台。穿过双闸门才能看到厚重的美式橡木门。总的来说，这是一栋坐落于洛杉矶最好的街道上、时尚而独特的住宅。

# Sands Point Residence

## 桑茨波恩特别墅

**Architects**
Narofsky Architecture

**Location**
Long Island, New York

**Photography**
Costas Picadas, Michael Grimm Photography

The residence started out as an idea to accommodate 3 generations of the same family. So it took its form, that the private spaces surrounded a great public space, which is the 2-storey central living, dining, entry and lounge space, a sort of indoor court yard. With its larges glazed areas this space as others in this home feel very attached to the outdoors. There is great continuity from indoor to outdoor also reinforced by using the same materials inside and out.

The house is mostly clad in a unique material, which is a 3 mm Slimtech porcelain panel manufactured by Lea Ceramiche from Italy. These panels some as large as 1m x 3m are mounted rain screen fashion (ventilated facade), on vertical insulated furring strips. The entry and the surfaces surrounding the main courtyard are clad in Zinc. All the roofs are green and are all accessible as usable as outdoor terraces.

/ 023

The house is slated to receive a LEED (Leadership in Energy and Environmental Design) for homes Silver as we incorporated many sustainable systems, such as: geothermal heating and cooling (fossil fuel is only used for cooking); LED lighting; water reclaim; super insulation (the entire house is foamed); steel frame (no wood was used); no maintenance materials like the Slimtech and zinc; as well as radiant heat throughout.

Other elements in the house are an indoor /outdoor pool area, with retractable doors for use in the warm weather seasons; an indoor squash/multi court; golf simulator/ theater room; two kitchens (one for cooking Indian food and isolate the spice odors); 7 full bedroom suites; billiard room, family room and a 5 car garage.

It all sits on 1.2 hectare on Long Island's North Shore, right on the Long Island Sound with beach access.

/ 035

该项目的出发点是建造一栋可以住下一家三代人的住宅。据此，便形成了一个私人空间围绕大家庭共用空间而设的布局。这个两倍高家庭公用空间内设有客厅、餐厅、门厅、休息空间，以及室内庭院。空间的四面大部分被玻璃包围着，给人一种里外相通的感觉。此外，室内室外采用了相同的建筑材料，增强了两个空间之间的延续性。

房子外面镶着 3mm 厚的 Slimtech 大瓷砖，这种特殊材料是由意大利 Lea Ceramiche 制造的，规格为 1m x 3 m，具有防水透气的特性，还带有竖向的隔热条。门厅和主庭院的外表被锌包裹着。房屋的屋顶是绿色的，而且可以上到屋顶上，很实用，就像户外露台一样。

房屋设计采用了许多可持续系统，旨在申请绿色家庭住宅银奖，例如地热供暖和制冷系统（矿物燃料仅供烹饪使用）、LED 照明系统、废水回收再利用系统、超级隔热设施（整栋别墅采用了具有泡沫结构的材料）、钢筋结构（未使用木材）、无需保养材料（Slimtech 瓷砖和锌）以及辐射热系统。

别墅设有室内外水池、暖季可拆除门、室内壁球多功能场地、高尔夫模拟空间（家庭影院）、两间厨房（其中一间用来烹饪印度菜，能防止香料的香味飘到其他空间）、7 套完整的卧室、台球室、家庭公共空间和一间 5 车位车库。

该住宅占地 1.2 公顷，位于长岛北海岸，恰好坐落在长岛海湾处，能直接通往沙滩。

# Harker Street, Plettenberg Bay

## 普利登堡湾哈克街豪宅

**Architects**
Greg Wright Architects

**Principal Architect**
Greg Scott

**Project Team**
Liana Abate, Mark Bardon

**Location**
Harker Street, Plettenberg Bay

**Site Area**
669 m²

**Building Footprint**
300 m²

**Gross Floor Area**
500 m²

**Photography**
Kate Del Fante Scott

Sited on an exceptionally steep piece of land on the bluff above BI Beach in Plettenberg Bay, the challenge with this project was always going to be how to make a beach house "live" across a vertically driven program. As a wonderful counterpoint to the challenges faced by the extreme topography, it was the steepness the site that offered the extensive, uninterrupted, panoramic views that extend beyond 180° vistas from Robberg Reserve in the south right around to Keurbooms lagoon in the north, the trick was how we balanced the two.

The program was unpacked in such a way that the living area filled the widest of the vertical platforms that was created so that kitchen, dining and living rooms as well as a guest bedroom could open onto a generous, partially covered terrace that made the most of the panorama described above. Pergolas, braai area, canopies level changes and the swimming pool offer a variety of "zones" that can be occupied in various manners depending on whether the terrace is used for a quiet spot of lounging or entertaining friends and family.

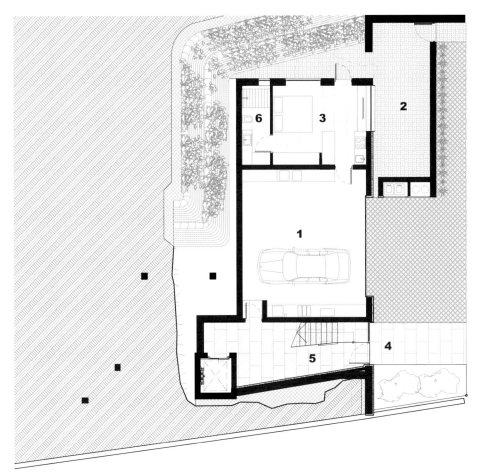

**LEGEND**

1) Double Garage
2) Courtyard
3) Domestic Quarters
4) Entrance
5) Foyer
6) En-suite
7) Family Room
8) Bedroom 1
9) Bedroom 2
10) Plant Room
11) Passage
12) Double Volume
13) Terrace
14) Pool
15) Lounge
16) Dining
17) Bedroom 3
18) Kitchen
19) Scullery
20) Guest WC
21) Main Bedroom
22) Dressing Room
23) Main Bathroom

The ground floor plays home to garages, staff quarters and a timber and stone clad entrance volume and stairwell; an intermediate level houses 2 other bedrooms and a small TV lounge whilst the top floor plays host to the master bedroom suite that completes the program.

Whilst unashamedly contemporary in the architectural language and interior detailing, a palette of timber and natural stone temper the building both inside and out. The materials are intended to weather into a series of greys and driftwood tones intended to ground the house in its coastal context.

Not only grey though, the house plays host to "green" components as well; the systems of the house are complemented by a series of sustainable features such as rainwater harvesting, PV panels with batteries amongst others.

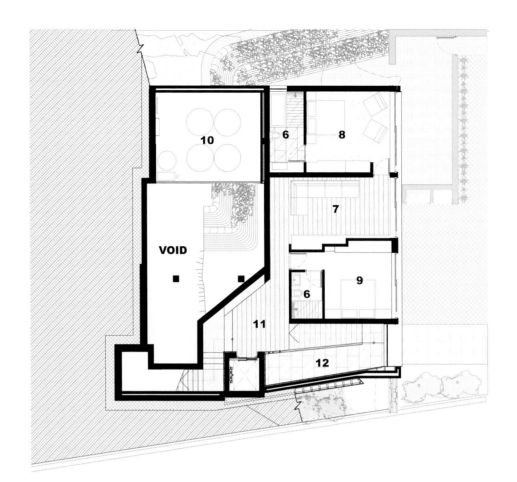

**LEGEND**

1) Double Garage
2) Courtyard
3) Domestic Quarters
4) Entrance
5) Foyer
6) En-suite
7) Family Room
8) Bedroom 1
9) Bedroom 2
10) Plant Room
11) Passage
12) Double Volume
13) Terrace
14) Pool
15) Lounge
16) Dining
17) Bedroom 3
18) Kitchen
19) Scullery
20) Guest WC
21) Main Bedroom
22) Dressing Room
23) Main Bathroom

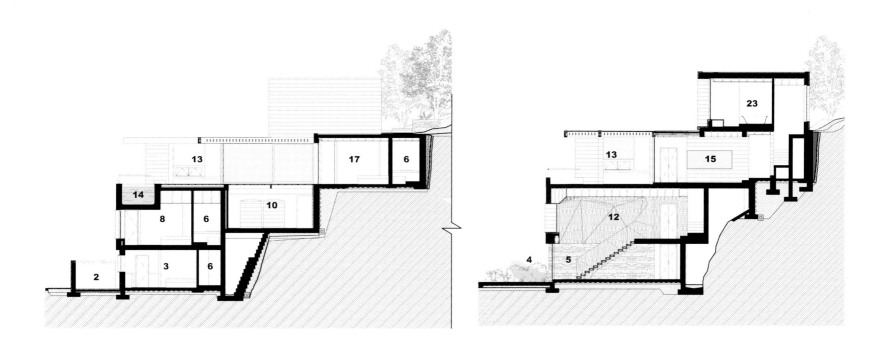

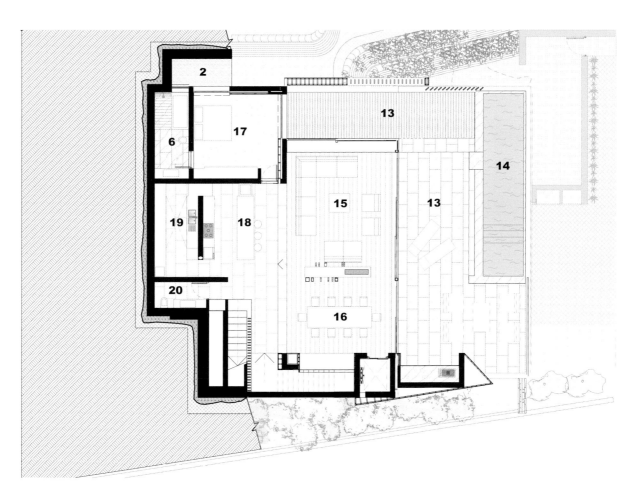

**LEGEND**

1) Double Garage
2) Courtyard
3) Domestic Quarters
4) Entrance
5) Foyer
6) En-suite
7) Family Room
8) Bedroom 1
9) Bedroom 2
10) Plant Room
11) Passage
12) Double Volume
13) Terrace
14) Pool
15) Lounge
16) Dining
17) Bedroom 3
18) Kitchen
19) Scullery
20) Guest WC
21) Main Bedroom
22) Dressing Room
23) Main Bathroom

　　该项目位于普利登堡湾BI海滩边的绝壁上，所处位置异常的陡峭。如何在纵向开凿的悬崖上建设一栋横向延伸的别墅一直是本项目所面临的挑战。极端地形带来了挑战，当然也带来了奇妙的观景视野，正是因为场地的陡峭，它才拥有了开阔、连续的全景式风光，如从南边的罗伯格自然保护区一直延续到北边的可鲁布环礁湖的180°远景，但重点是如何实现两者之间的平衡。

　　项目的起点是把生活区设在最宽的垂直平台上。正是如此，厨房、餐厅、客厅和客房都朝向部分被覆盖的大型露台，从而保证了每个空间的风景。凉亭、烧烤区和顶篷的水平高度都被改变了。游泳池则提供了各种功能分区，这取决于露台是用作安静的休息区，还是用来款待朋友和家人。

　　一楼设有车库、员工宿舍、楼梯间以及被木材和石头覆盖的大门。中间层设有2间卧室和小型电视间，而顶层则是主卧室套房。

　　此外，建筑具有名副其实的当代建筑语言和室内设计细节，木材和天然石材是建筑的基调，实现了内外的协调。为了让房子融入海岸环境，建筑材料都故意被风化为灰色调，并被设计成浮木风格。

　　别墅不仅采用了灰色调，还加入了绿色元素。一系列的可持续设施完善了别墅的系统，如集雨设备、自带蓄电池太阳能板等。

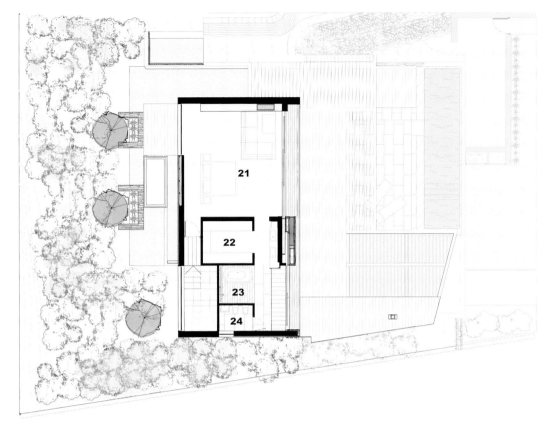

**LEGEND**

1) Double Garage
2) Courtyard
3) Domestic Quarters
4) Entrance
5) Foyer
6) En-suite
7) Family Room
8) Bedroom 1
9) Bedroom 2
10) Plant Room
11) Passage
12) Double Volume
13) Terrace
14) Pool
15) Lounge
16) Dining
17) Bedroom 3
18) Kitchen
19) Scullery
20) Guest WC
21) Main Bedroom
22) Dressing Room
23) Main Bathroom

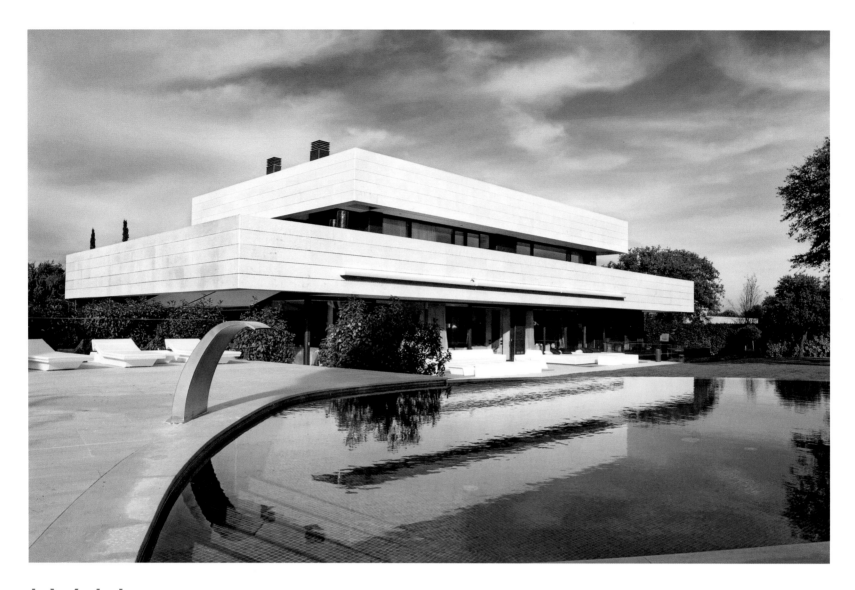

# LV House

## LV 屋

**Architects**
A-cero, Joaquín Torres &
Rafael Llamazares architects

**Location**
Spain

**Photography**
Plasmalia

A-cero presents a new single-family house located in the north of Spain. This house with an approximate built up area of 1,000 m² is divided in three floors. The project is characterized by its simplicity and its blend of classic style with modern flair. The property is located on a large landscaped garden and a swimming pool with organic shapes.

With a different aesthetic criteria that characterize the studio managed by architects Joaquín Torres and Rafael Llamazares. This house designed by A-cero is provided outdoors and indoors with high standard quality materials and furniture.

When people get into the house through the distributor hall they get surprised by the amazing staircase that connects the three floors. Downstairs, in the ground floor they find the public areas like dining and living rooms as well as the kitchen and the service area. All the rooms have been designed with wide windows. All the rooms are in connection with the garden through the porches.

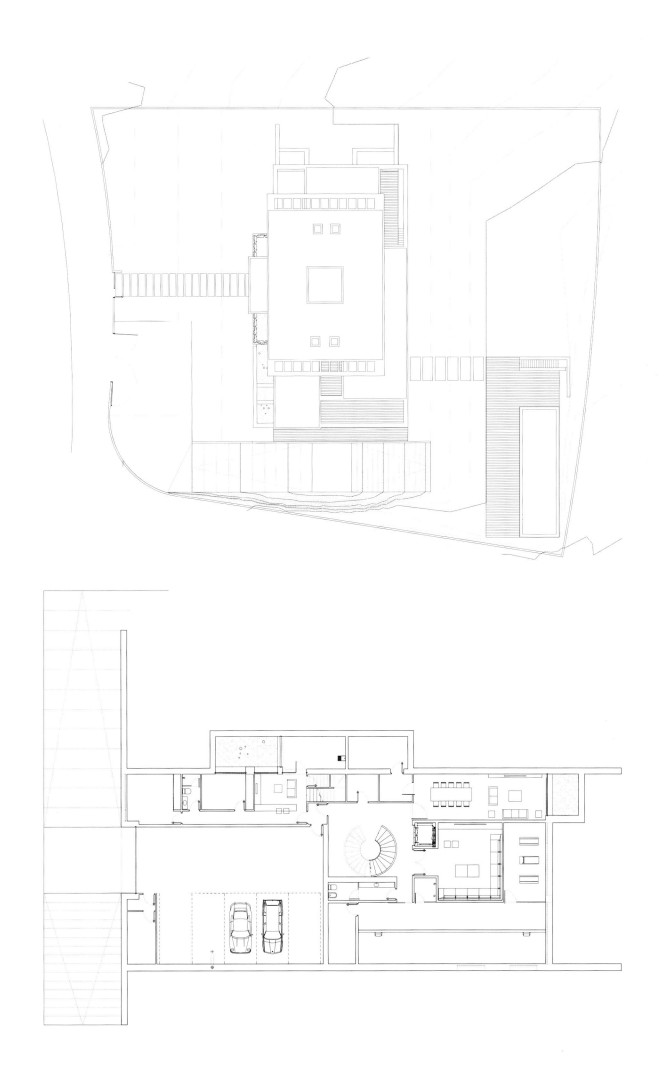

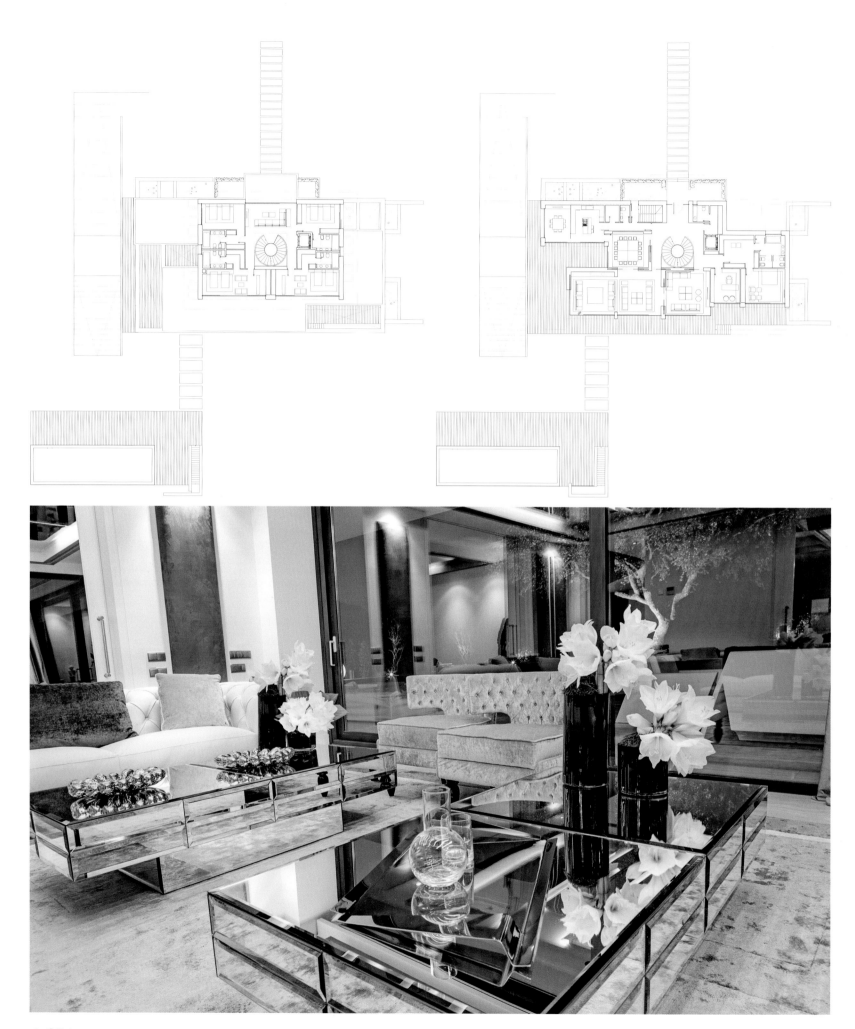

/ 055

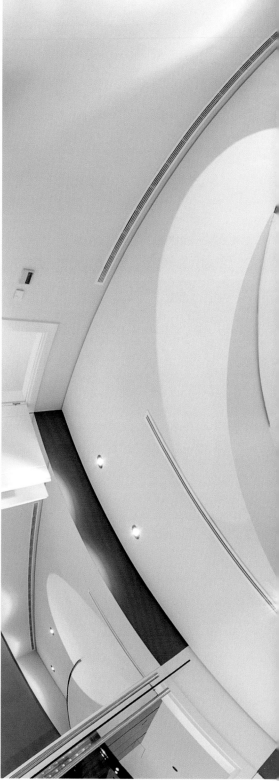

The top floor is reserved for the bedrooms. There is a master bedroom with bathroom and dressing room and 4 other bedrooms also with its own bathroom.

In the basement they find the garage, an entertainment area, which is perfect to meet people and also the indoor pool with gym. The property also has an elevator to connect the different floors.

The interior design and the furniture chosen by the client have different origins. There are several Ipe Cavalli furniture designs as well as A-cero IN designs. Vondom firms the outdoor furniture. The decoration is completed with different objects from the client and also furniture from A-cero IN like sculptures designed by the architectural studio and objects from antique shops or stores like Anmoder and Lou & Hernandez.

A-cero 设计的新作 LV 屋位于西班牙北部，属于独户住宅。别墅占地约 1 000 m²，建有 3 层楼。该项目以简约为特色，融合了古典风格和现代风格。别墅坐落在大景观花园中，其中的游泳池也具有有机的形状。

由杰昆·托雷斯（Joaquín Torres）和拉斐尔·利亚马萨雷斯（Rafael Llamazares）两名设计师负责管理的 A-cero 工作室，与众不同的审美标准是该工作室的特色。在这栋别墅中，高质量的材料和家具都被应用到室外和室内空间中。

步入大厅，映入眼帘的便是通往各层的楼梯。楼下为设有餐厅、起居室、厨房和服务区域的一楼空间。此外，所有的空间内都设有大窗户，通过门廊与花园相连。

顶层的空间是卧室。主卧室配有浴室和更衣间，其他 4 个卧室也都有独立浴室。

一楼空间设有停车库和适合聊天的娱乐空间，还有室内游泳池和健身房。别墅还装有电梯，方便通往各层。

由客户自主选定的室内设计和家具都有着不同的来源。其中含有 Ipe Cavalli 和 A-cero IN 设计的家具，还有 Vondom 公司的户外家具。房屋装饰采用了客户自己的各种物品和 A-cero IN 的家具，例如由建筑公司设计的雕塑，以及从古董店、Anmoder 和 Lou & Hernandez 等商店选来的物品。

/ 061

# San Vicente Residence

圣文森特别墅

**Designer**
Paul McClean

**Location**
Los Angeles, CA

**Photography**
Jim Bartsch

The new residence is located on a large flat lot adjacent to a busy street. The designers' goal was to design a beautiful family home while minimizing street noise. It is essential that the outside world be left behind and their idea focused on creating a zone of decompression as you enter the property. The street is heavily screened with landscaping and high walls. The drive court is separated from the entry courtyard by a glass screen wall. The landscaped courtyard contains a water element to further screen noise and provide an attractive focal point. The house is laid out over 2 levels in an "L" shape configuration around the garden. The kitchen family room opens onto a large covered porch which is lit through a glass floor in the deck above. The pool house provides an attractive focal point from the living spaces across the garden. There are 6 bedrooms in the house as well as formal and family living spaces, media and office. The palette of materials is soft contemporary with extensive use of limestone, wood as well as Italian cabinetry and bronze accents. The garden is the focus of the design incorporating the lawn and pool, various outdoor entertaining spaces, ornamental trees and sculpture.

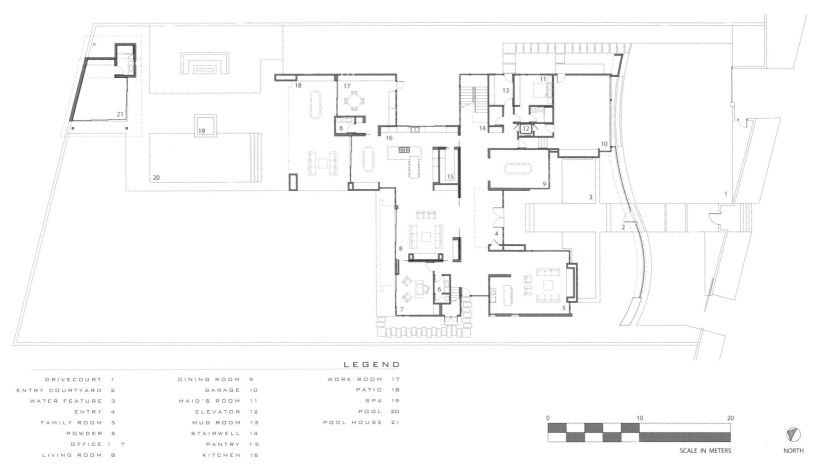

LEGEND

| | | |
|---|---|---|
| DRIVECOURT 1 | DINING ROOM 9 | WORK ROOM 17 |
| ENTRY COURTYARD 2 | GARAGE 10 | PATIO 18 |
| WATER FEATURE 3 | MAID'S ROOM 11 | SPA 19 |
| ENTRY 4 | ELEVATOR 12 | POOL 20 |
| FAMILY ROOM 5 | MUD ROOM 13 | POOL HOUSE 21 |
| POWDER 6 | STAIRWELL 14 | |
| OFFICE 1  7 | PANTRY 15 | |
| LIVING ROOM 8 | KITCHEN 16 | |

MAIN LEVEL PLAN

/063

　　该新住宅临近繁华的街道，场地平坦宽阔。设计旨在打造一栋美观的家庭住宅，同时减少从街上传来的噪音，重点则是让人遗忘外部世界，而内部空间又可以拂去人的压力。景观和高墙就像屏风，遮住了街道。玻璃幕墙把车道从入口庭院隔离开来。庭院中的水景进一步隔离了噪音，成为一道引人注目的风景。建筑中不止有两个层次是围绕着花园呈"L"形展开的。厨房客厅空间敞向宽阔的走廊，走廊设在延伸出来的屋檐下，灯光透过露天平台上方的玻璃表面，照亮着走廊。从客厅往花园看去，视线会聚焦在池边的小屋上。屋内有6间卧室、正式客厅和家庭休息客厅、媒体间和办公室。建筑大量采用了石灰岩、木材、意大利橱柜和铜器等当代软装饰材料。花园是设计的重点，草坪与水池的结合提供了充足的户外娱乐空间，观赏树木和雕塑更是美化了花园。

### LEGEND

| | | |
|---|---|---|
| GYM 1 | BEDROOM 2 9 | M. CLOSET 17 |
| STUDY 2 | CLOSET 2 10 | M. BATHROOM 18 |
| LAUNDRY 3 | BATHROOM 2 11 | DECK 19 |
| GUEST BEDRROM 4 | BEDROOM 3 12 | |
| FAMILY ROOM 5 | BATHROOM 3 13 | |
| BEDROOM 1 6 | CLOSET 3 14 | |
| CLOSET 1 7 | LOUNGE 15 | |
| BATHROOM 1 8 | MASTER BEDRROM 16 | |

SCALE IN METERS

NORTH

UPPER LEVEL PLAN

/ 073

# Sunset Strip Residence

日落大道豪宅

**Designer**
Paul McClean

**Location**
Sunset Strip, Los Angeles

**Photography**
Jim Bartsch

The house occupies a hilltop lot in one of the most desirable locations in Los Angeles. It is long and narrow and the primary view was only available from one end of the lot. The designers' solution was to create a compound of three buildings; the main house, garage and wellness as well as a guest house. Passing through the gate, a landscapes hedge leads to the drive court which is centrally located between the three buildings. The garage sits in a spot that enjoys spectacular views of the surrounding canyons so they designed it to be glazed on both sides. All three buildings are connected by a water feature that leads the eye towards the views and entry. The main house is approached along the water feature. The front hallway, glazed and open on three sides, leads to a stairwell where a beautiful chrome and stone stair ascends to the upper level bedrooms.

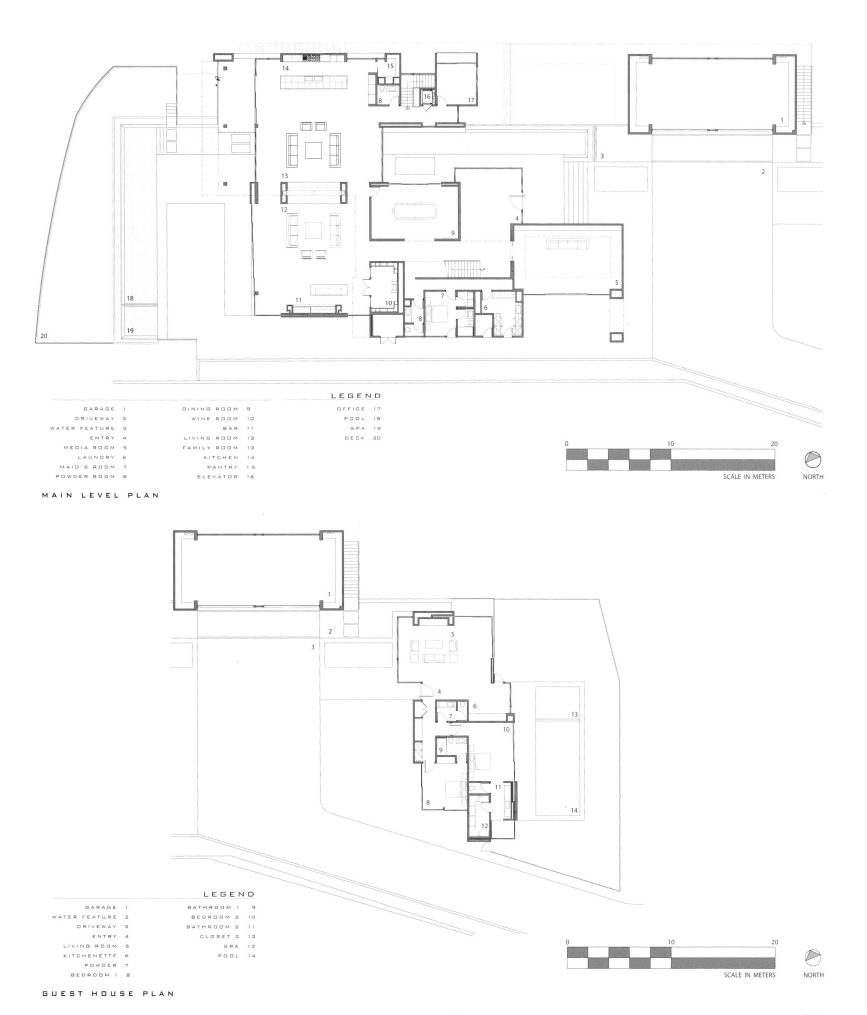

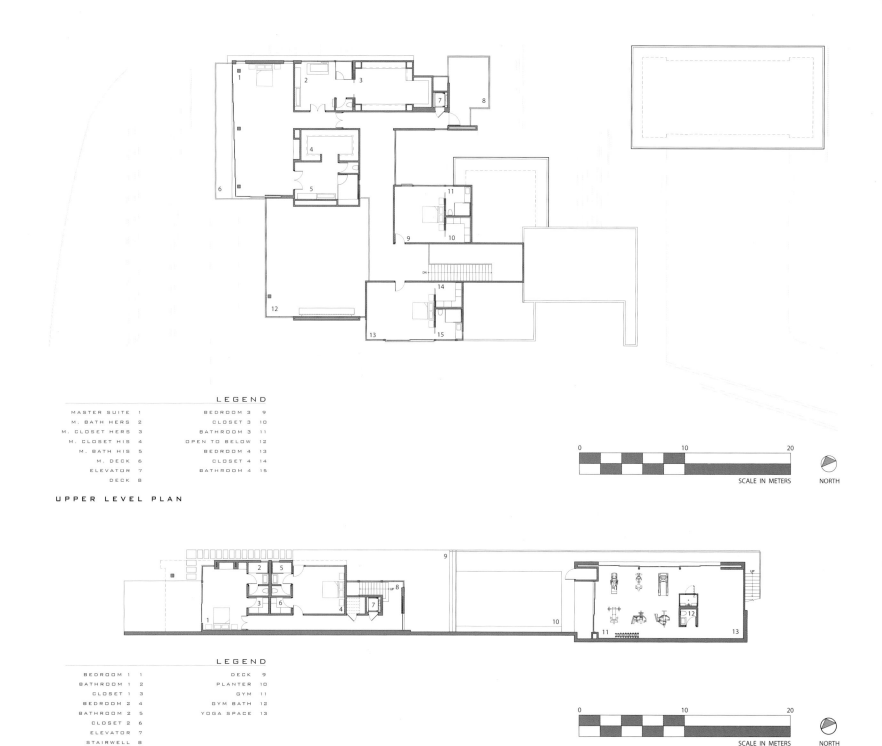

LEGEND

| | | | |
|---|---|---|---|
| MASTER SUITE | 1 | BEDROOM 3 | 9 |
| M. BATH HERS | 2 | CLOSET 3 | 10 |
| M. CLOSET HERS | 3 | BATHROOM 3 | 11 |
| M. CLOSET HIS | 4 | OPEN TO BELOW | 12 |
| M. BATH HIS | 5 | BEDROOM 4 | 13 |
| M. DECK | 6 | CLOSET 4 | 14 |
| ELEVATOR | 7 | BATHROOM 4 | 15 |
| DECK | 8 | | |

UPPER LEVEL PLAN

LEGEND

| | | | |
|---|---|---|---|
| BEDROOM 1 | 1 | DECK | 9 |
| BATHROOM 1 | 2 | PLANTER | 10 |
| CLOSET 1 | 3 | GYM | 11 |
| BEDROOM 2 | 4 | GYM BATH | 12 |
| BATHROOM 2 | 5 | YOGA SPACE | 13 |
| CLOSET 2 | 6 | | |
| ELEVATOR | 7 | | |
| STAIRWELL | 8 | | |

LOWER LEVEL PLAN

The main living room is 2-storey tall and enjoys spectacular views of the Los Angeles Basin and the ocean beyond. The room incorporates a bar and glazed wine cellar as well as an elongated see through fire place that is visible from the family room on the other side. The combined kitchen and family room has a more intimate feeling than the living room and appears to float over the water feature. From here it is possible to look back along the water past the garage all the way to the guest house and beyond. This level of the house is completed by service spaces and an office for the owner. The upper level contains the master with his and her baths and closets as well as 2 secondary bedrooms. 2 further bedrooms are located in the basement as well as the guest house across the drive court. The palette of materials is soft contemporary with extensive use of limestone, wood as well as Italian cabinetry and bronze accents.

该住宅坐落在小山的山顶上，是洛杉矶最适宜居住的位置，从该场地的一端能欣赏到此处主要的远景。设计方案是打造3栋复合建筑，主楼设立车库、健身房和客房。穿过大门，树篱小道直接通往铺在3栋楼房中央的车道。车库所处位置优越，能欣赏到壮丽的峡谷风景，因此其两侧都采用了玻璃墙。水景将3栋楼房连接起来，将人的视线引向风景和入口。主楼沿水景延伸，前廊的3面都是玻璃，通往用铬和石材打造的楼梯，穿过楼梯，便来到了顶层的卧室。

两倍高的主客厅拥有迷人的洛杉矶盆地风光和远处海洋景观，客厅内设有吧台、玻璃酒窖和壁炉。另一端的家庭休息室也设有加长的透明壁炉。厨房和家庭休息组合空间仿佛悬浮在水景上方，氛围比客厅更温馨亲切，从此处往回看，能看到水景和车库，还有客房，甚至更远处，视线毫无阻碍。住宅的该层设有服务空间和主人的办公室。顶层则是主卧室、男女浴室、衣柜和两间次卧。还有两间卧室设在地下室和车道对面的客房。石灰石、木材、意大利橱柜和铜器等当代软装饰材料充当了建筑的主材料。

# Butterfly House

---

蝴蝶屋

**Architects**
jmA

**Principal Architect**
John Maniscalco

**Project Architect**
Kelton Dissel

**Project Team**
John Maniscalco, Kelton Dissel, Marit Gamberg

**Interior Furnishings**
Shawback Design

**Location**
San Francisco, CA

**Area**
462 m²

**Photography**
Joe Fletcher

In this complete rebuild of a mid-century modern home, the design flows from an analysis of the varied site conditions already present and reinforces key relationships to the site while establishing new ones. From a relocated street level entry, a careful sequence follows the slope and curates the vertical movement through the home from earthbound experience to the open sky and panoramic views. Each level takes on a different purpose first, establishing new ties to the street, then anchoring family spaces to the south-facing garden, turning inwardly focused at the sleeping level, and ultimately dissolving at the top level living spaces and roof deck to reveal panoramic connections to the city and bay.

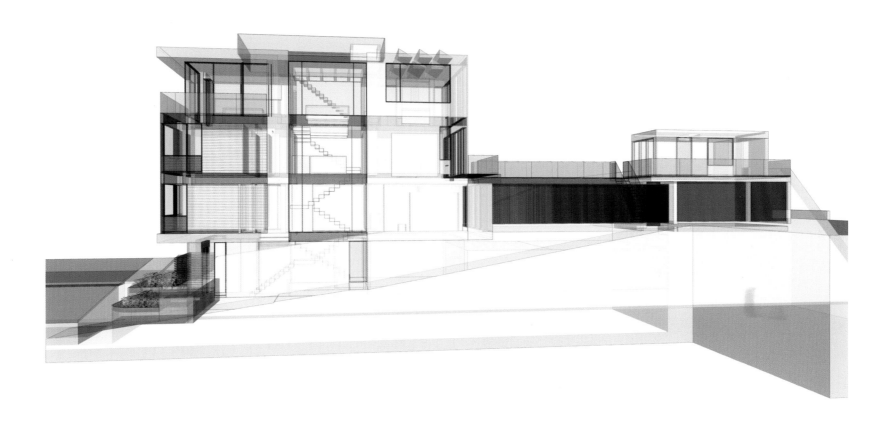

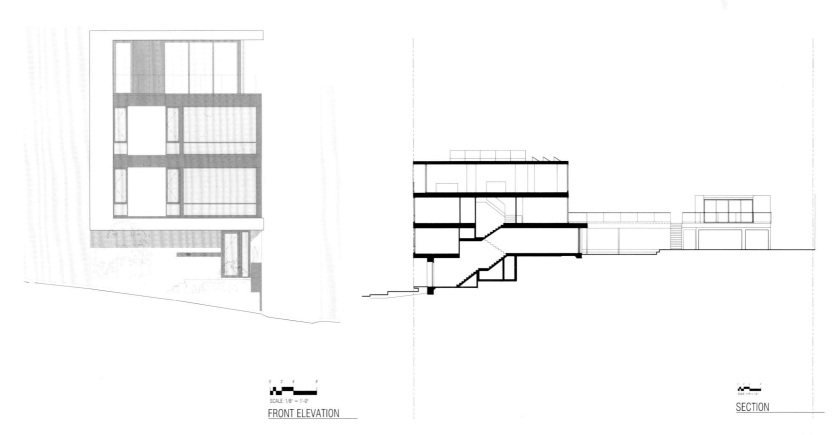

SCALE: 1/8" = 1'-0"
FRONT ELEVATION

SECTION

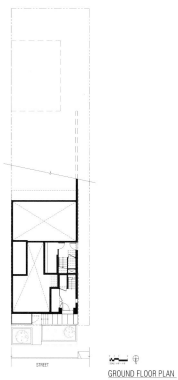

GROUND FLOOR PLAN

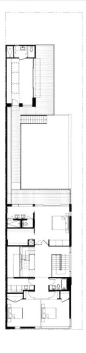
FIRST FLOOR PLAN

SECOND FLOOR PLAN

/ 087

THIRD FLOOR PLAN

ROOF PLAN

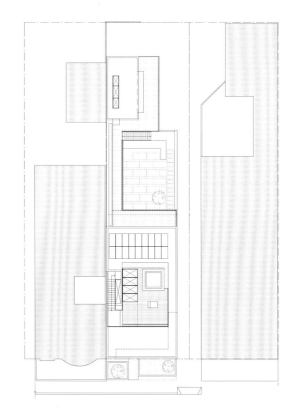

SITE PLAN

/ 088

这栋现代建筑建于21世纪中叶，在其翻修和设计过程中，首先分析了现存场地的条件。在与场地建立新关系的同时，也突出了建筑与场地之间已有的重要关系。重新选址的入口与街面同高，依照斜坡的坡度谨慎地排列着，让屋内的仅限于地面的空间体验垂直转移到了开阔的天空，引入无限的风景。设计中每个层次的目标都不一样，先是与街道建立新联系，然后把家庭空间定位在朝南的花园中，再转向内部，关注休息空间，最后是顶层的生活空间和露台，并引入整个城市和海湾的风光。

# Cliff House

崖居

**Architects**
SAOTA

**Architects Project Team**
Stefan Antoni, Greg Truen
& Juliet Kavishe

**Interior Design**
Antoni Associates

**Interior Design Team**
Mark Rielly & Sarika Jacobs

**Main Furniture Supplier**
OKHA Interiors

**Building Area**
1,954m²

**Location**
Dakar, Senegal

**Photography**
SAOTA

The ground floor of the house, designed to facilitate seamless indoor and outdoor living and entertainment, is arranged in an "L" shape around the pool, the pool terrace and the garden. The formal living and dining spaces cantilever over the cliff and hang over the Atlantic Ocean enjoying panoramic sea views as well as views back to the house. The kitchen made up of a so called "American" or open kitchen and a separate traditional kitchen as well as the garage and staff facilities run along the east west axis and along the northern side of the boundary. From the entrance one moves past the sculptural circular stair to the entertainment room and the double volume family lounge which connects up with a floating stair to the upper level pyjama lounge. The main and the two children's bedrooms are placed on this upper level.

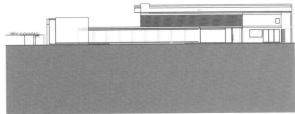
NORTH EAST ELEVATION

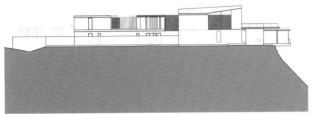
NORTH WEST ELEVATION

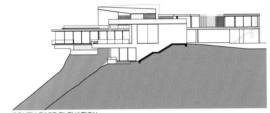
SOUTH EAST ELEVATION

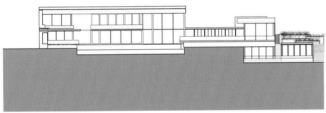
SOUTH WEST ELEVATION

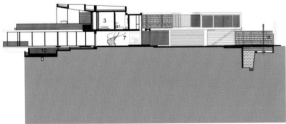
SECTION A-A

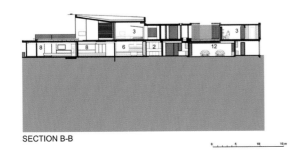
SECTION B-B

LEGEND
1. CINEMA
2. SERVICE AREA
3. BEDROOMS
4. POND
5. GYM
6. KITCHEN
7. ENTRANCE
8. LOUNGE
9. STUDY
10. POOL
11. TERRACE
12. GARAGE
13. ENTRANCE GATE
14. GATE HOUSE

"The huge overhanging roof which projects over the upper level and the outdoor living level creates a dramatic double volume outdoor space and gives the entire home a sense of unity." says partner Stefan Antoni.

One of the features of the house is the spiral staircase, clad in stainless steel, while the treads are clad in white granite. To add to the sense of continuity between the levels the 20 mm in diameter stainless steel rods run from the first floor handrail to the lower ground floor, thus making the stairwell look like a sculptural steel cylinder. A skylight above the stairwell as well as floor to ceiling glazing in the lounges adds to the sense of transparency.

The main bedroom suite opens up onto a large terrace which is the roof of the more formal living wing of the house and the element which projects over to the ocean. The main bathroom opens into a private garden and outdoor shower situated over the garages.

The study / office sits in a separate block and is joined to the main house by a hallway running along the spine of the building. Under the study/office is a separate fully contained guest room, alongside which is a private gym and reflecting pond.

　　该住宅的第一层沿水池、水池露台和花园呈"L"形展开，其设计旨在将室内空间和户外的起居和娱乐空间连为一体。主客厅和餐厅伸出悬崖，悬在大西洋上，不但可以俯瞰广阔的海景，还能观赏屋后的花园。厨房融合了美式（开放式）厨房和传统独立厨房两种风格，车库和员工房屋沿北边的界线自东向西而建。由大门进入屋内，再沿着庞大的螺旋式楼梯，就可进入娱乐空间和两倍高的家庭休息间。该空间中设有悬浮式楼梯，能进入到楼上的休闲休息室。主卧室和两间儿童房也都设在顶层。

　　"顶层大屋顶外悬出来的屋檐与露天起居空间层一起，形成巨大的两倍高室外空间，赋予整栋房屋一种整体感。"合作人斯蒂芬·安东尼说。

　　房屋的特色之一就是螺旋式楼梯，其材料选用了不锈钢，踏板采用了白色花岗岩。20mm粗的不锈钢扶手从首层延伸至一楼，增加了楼梯的延续性，使楼梯变成了一个巨大的钢滚筒。楼梯间的顶端是天窗，与休息室的玻璃幕墙相呼应，给人以一种通透感。

　　主套房敞向大露台，露台又是悬在海洋之上的主客厅的屋顶。主浴室设在车库上方，敞向私人花园和露天淋浴空间。

　　书房（办公室）设在独立的区域，通过建筑中央的走廊与主建筑相连。书房下方是完全独立的客房，客房旁边则是私人健身房和清澈的水池。

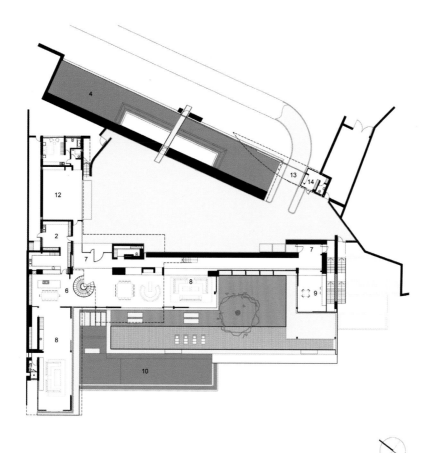

GROUND FLOOR

LEGEND
1. CINEMA
2. SERVICE AREA
3. BEDROOMS
4. POND
5. GYM
6. KITCHEN
7. ENTRANCE
8. LOUNGE
9. STUDY
10. POOL
11. TERRACE
12. GARAGE
13. ENTRANCE GATE
14. GATE HOUSE

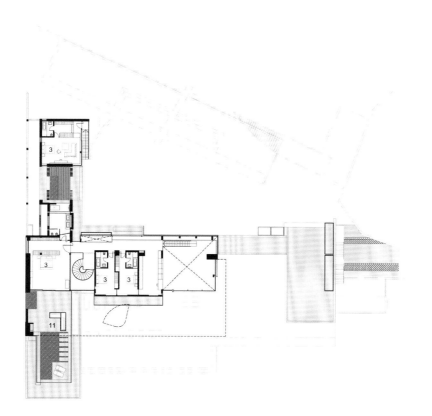

FIRST FLOOR

LEGEND
1. CINEMA
2. SERVICE AREA
3. BEDROOMS
4. POND
5. GYM
6. KITCHEN
7. ENTRANCE
8. LOUNGE
9. STUDY
10. POOL
11. TERRACE
12. GARAGE
13. ENTRANCE GATE
14. GATE HOUSE

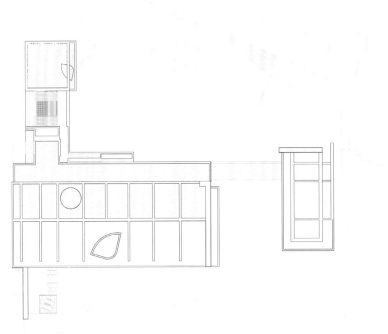

ROOF PLAN

**LEGEND**

1. CINEMA
2. SERVICE AREA
3. BEDROOMS
4. POND
5. GYM
6. KITCHEN
7. ENTRANCE
8. LOUNGE
9. STUDY
10. POOL
11. TERRACE
12. GARAGE
13. ENTRANCE GATE
14. GATE HOUSE

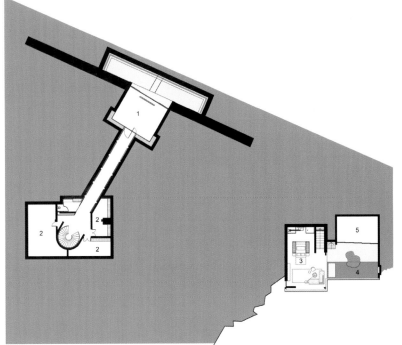

LOWER GROUND

**LEGEND**

1. CINEMA
2. SERVICE AREA
3. BEDROOMS
4. POND
5. GYM
6. KITCHEN
7. ENTRANCE
8. LOUNGE
9. STUDY
10. POOL
11. TERRACE
12. GARAGE
13. ENTRANCE GATE
14. GATE HOUSE

/ 099

# Cove 3

海湾三号

**Architects**
SAOTA

**Architects Project Team**
Greg Truen & Roxanne Kaye

**Interior Design**
Antoni Associates

**Interior Design Team**
Mark Rielly & Tavia Pharaoh

**Building Area**
1,005 m²

**Location**
The Cove, Pezula Estate, Knysna

**Photography**
SAOTA, John Devonport, Adam Letch

The primary idea driving the design was to create a single living space with a single roof element floating over it that responded to the slope of the site. The roof is set at a sufficiently high level so that it is out of one's line of sight from the living space, creating the illusion that one is sitting in the landscape rather than in a room looking out into a landscape.

A large triangular cut-out in the roof reinforces a connection with the sky. A very detailed solar analysis was done of the building to try and get direct sun (other than the rising east sun) out of the building. As a result, a midlevel horizontal sunscreen was added to the double height glass facade and the skylight is protected by a timber screen that hangs into the space to mitigate the scale of the double volume space. Care was also taken in selecting performance-glass that would minimise the impact of direct sun.

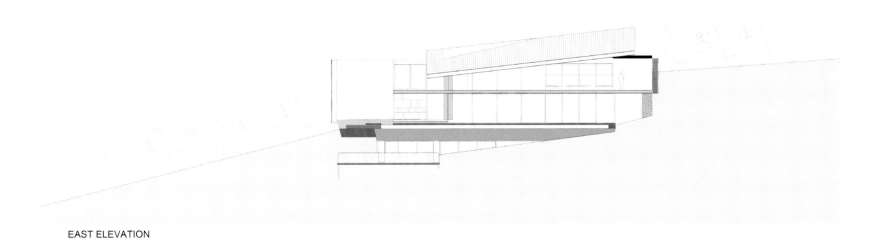

EAST ELEVATION

SECTION A-A

LEGEND
1. ENTRANCE
2. BAR
3. KITCHEN
4. BEDROOM
5. DINING
6. POND
7. LOUNGE
8. DECK / TERRACE
9. POOL
10. GARAGE
11. STUDY
12. LIFT
13. HOME CINEMA

The building is approached from the north west at the top of the site. The choice of materials, off-shutter concrete, Rheinzink roofing, timber cladding, stone and exposed aggregate will allow the building to fade into the landscape as it ages. The building is orientated towards the view; one enters at the upper level of the double volume looking towards the ocean. The contrast with the external approach is very powerful. A grand stair pulls on onto the living level which holds the kitchen, dining room and living room. To the right the landscaping is pulled into the building, blurring the distinction between the inside and the outside.

A spiral stair connects the living level to a private lounge and the master bedroom on a mezzanine level. This stair was conceived as a sculptural element in the large volume to again mitigate the scale of this space. This spiral drops through the floor to a lower level which houses a guest bedroom, a home theatre and a living room. An "L" shaped extension to the south west houses the two children's bedrooms. The bedrooms have curved curtain tracks that create very intimate sleeping spaces at night which contrast with the very open daytime character.

Water is a critical issue in this part of the world and a huge underground cistern was created under the garden terrace to harvest rainwater to minimise the houses' reliance on the municipal water system. A heat pump and water based under floor heating system uses less energy than would ordinarily be required for a house of this magnitude. The concept behind the landscaping was to reinstate the fynbos and let the building float over this restored surface.

设计师的构想是创建一个与住宅斜坡选址相呼应的生活空间，住宅屋顶高高地悬浮在生活空间的上面，远远望去如同被置于周边山地树木的簇拥中，给人以一种生活空间被隐藏的错觉。

屋顶上的三角形大开口朝向天空。通过对建筑位置与日照方向进行仔细分析，使房屋避开太阳光的直射（而非太阳初升时的阳光）。两倍高玻璃幕墙的中间横向贴着遮阳膜，悬于此空间的木遮板挡住了天窗处的阳光，也缓和了空间的空旷感。精心挑选的高效能玻璃将阳光直射对房屋的损害减到了最小。

住宅的入口设在西北方向，也是海崖的最高处。建筑是由混凝土外墙、莱茵锌屋顶、木材包层以及裸露的石头组成，随着时间的推移，房屋能够更好地融入到周围景观中。别墅敞向壮丽的风光，进入两倍敞高空间的上层空间，就能看到浩渺的海洋美景，与室内空间形成强烈的对比。沿大楼梯走到起居空间，可以看到都设在此空间中的厨房、餐厅和客厅。从空间的右侧可以看到美丽的风景，这便模糊了室内与室外空间的界线。

大螺旋式楼梯连接了私人休息空间和中间层的卧室，就像是屹立在偌大空间中的雕塑，进一步缓和了这个大空间。楼梯穿透地面，延伸到设有客房、家庭影院和客厅的更低一层；延伸至西南方向的"L"形空间是两位儿童的卧室。卧室中的窗帘可以沿弯曲的滑道拉升，使得空间在夜间具有温馨的气氛，而白天又显得开阔。

在当地，水是关键问题。为了打造一幢环保可持续的的住宅，设计师在花园露台的位置设计了一个巨大的地下蓄水池用来收集雨水。另外，地热供暖系统能满足别墅日常的暖气和热水供应。对于景观，其设计理念是恢复高山硬叶灌木群，营造一种建筑飘浮在这景观之中的感觉。

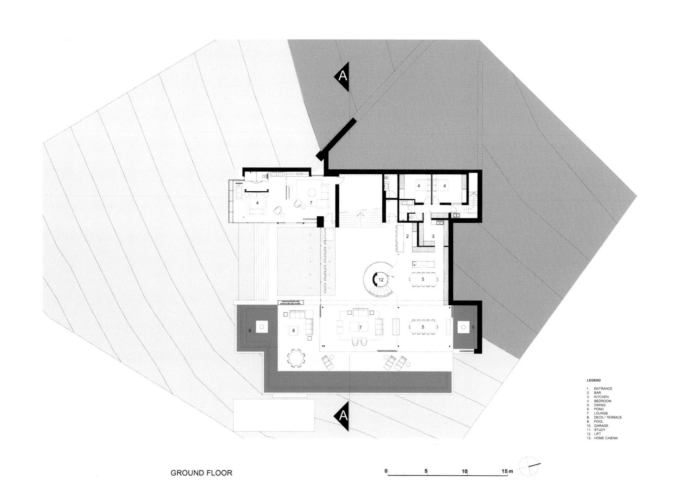

GROUND FLOOR

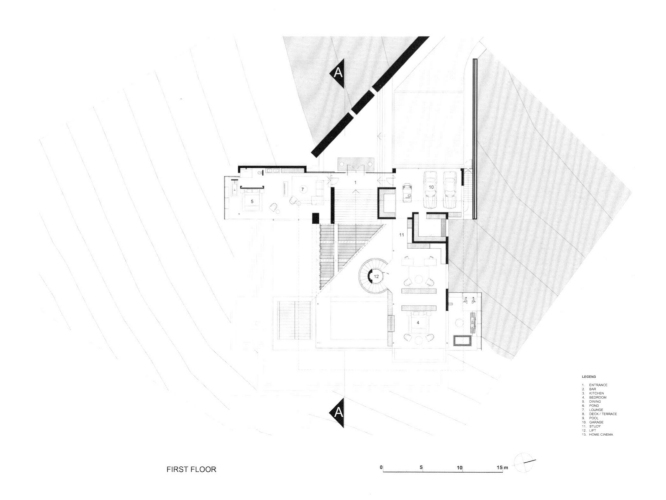

FIRST FLOOR

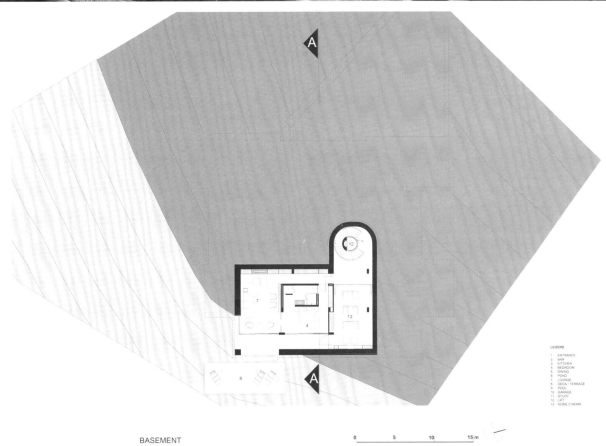

BASEMENT

LEGEND

1. ENTRANCE
2. BAR
3. KITCHEN
4. BEDROOM
5. DINING
6. POND
7. LOUNGE
8. DECK / TERRACE
9. POOL
10. GARAGE
11. STUDY
12. LIFT
13. HOME CINEMA

# Head Road 1818

## 首脑路 1818 号别墅

**Architects**
SAOTA

**Project Team**
Philip Olmesdahl, Stefan Antoni
& Mark Bullivant

**Interior Decor**
Craig Kaplan

**Location**
Fresnaye, Cape Town

**Photography**
Adam Letch

The clients acquired a steeply sloping site on Head Road and wanted to capitalize on the property in line with their lifestyle. Their focus was predominantly on a home that lived well on the site, with one great master suite, a number of guest rooms as well as ancillary rooms.

Whilst the site enjoyed great views to the ocean and back up to the mountain, Head Road has its own unique zoning which prescribed a large lateral boundary set back, and a height restriction aligned to the steep slope of the road. In addition to these restrictions, both lateral neighbours had built very large dwellings which towered over the narrow property. Fortunately due to the steep slope of the site, there was no

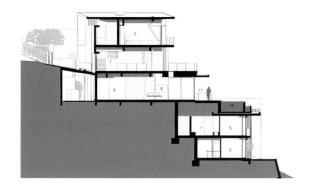

**SECTION AA**
**HEAD 1818**
1 : 200

LEGEND
1. MAIN ENTRANCE
2. GARAGE
3. STUDY
4. LIFT
5. BEDROOM SUITE
6. DRESSING
7. DINING ROOM
8. LOUNGE
9. KITCHEN
10. DOUBLE VOLUME
11. TERRACE
12. POOL
13. BAR
14. LAUNDRY
15. GYM

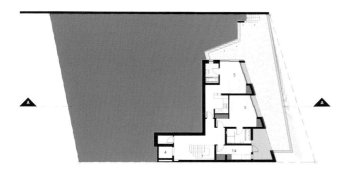

**LOWER BASEMENT LAYOUT**
**HEAD 1818**
1 : 200

LEGEND
1. MAIN ENTRANCE
2. GARAGE
3. STUDY
4. LIFT
5. BEDROOM SUITE
6. DRESSING
7. DINING ROOM
8. LOUNGE
9. KITCHEN
10. DOUBLE VOLUME
11. TERRACE
12. POOL
13. BAR
14. LAUNDRY
15. GYM

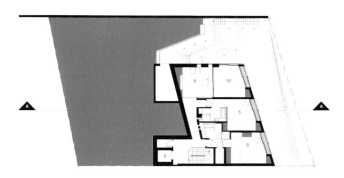

**BASEMENT LAYOUT**
**HEAD 1818**
1 : 200

LEGEND
1. MAIN ENTRANCE
2. GARAGE
3. STUDY
4. LIFT
5. BEDROOM SUITE
6. DRESSING
7. DINING ROOM
8. LOUNGE
9. KITCHEN
10. DOUBLE VOLUME
11. TERRACE
12. POOL
13. BAR
14. LAUNDRY
15. GYM

negative influence on the house's outlook.

Effort was put into capitalising on views to the rear onto Lion's Head as well as across sea point and the Atlantic Seaboard. Such interesting dynamics impacted on the placement of accommodation, with the required levels of privacy. There is a clear vertical distinction between the living and bedroom accommodation. The upper-most level accommodates the master bedroom; a generous open-plan space inclusive of a dressing area and en-suite bathroom, which enjoys bidirectional views (i.e. a sea-facing terrace and raised clear storey views towards Lion's Head).

From the entrance, a timber staircase accesses the primary living and entertainment area through a double volume space.

This open-plan arrangement of dining, lounge, kitchen and breakfast area offers seamless connection to the terraces, garden and pool. The various functions within this versatile space are subtly defined through the ceiling plane, sliding doors which open dramatically.

The language of the house reflects bold contemporary lines and extended.

Challenges were turned into opportunities where screening elements ensuring privacy have become an integral component of the aesthetic. The juxtaposition of solid mass to large expanses of transparency is heightened due to the choice of materials and introduction of perforated panels. Emphasis was placed on a modern, contemporary aesthetic with a high level of comfort. The architectural language is bold and muscular with powerful, almost graphic facade layouts with glazed voids and planar walls that open up to create generous ocean views.

The angular nature of the site resulted in a number of complex junctions which were embraced, to emphasise the angularity of certain elements while capitalising on the valuable land. The tactile qualities of timber and the way it is used activate the senses — the sound and smell of the hardwood timber underfoot on the staircase sets a tone on entering the house. This is a remarkable site with amazing views; and it was critical that as large a living area was created with strong connections to the outside living spaces. Such a narrow, steep site often compromises the final outcome but here, with the seamless connection of indoor and outdoor spaces, the heart of the house was successfully extended from boundary to boundary.

The house has a clearly defined upper and lower portion. The upper three levels of the home consists of the living room / terraces level, topped by the entrance / garage level and the upper-most level is the master suite. The lower portion is two storeys, accommodating the pool structure, multiple guest rooms, staff quarters and plant rooms. All levels are connected by a lift.

Landscaping plays a key part of the design, with the landscaping undertaken by Franchesca Watson and characterised by architectural, robust planting. The scale of the various planted areas is varied through the scaling device of an interesting aluminium pergola. Planted areas add softness to the entrance area, and the pool terrace, as well as contribute to screening houses on the lateral boundaries as the house terraces down the property.

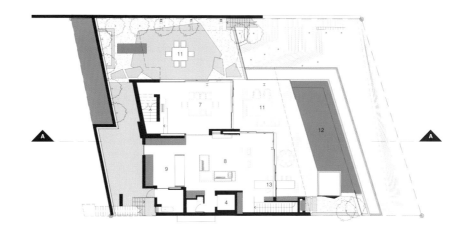

**LEGEND**

1. MAIN ENTRANCE
2. GARAGE
3. STUDY
4. LIFT
5. BEDROOM SUITE
6. DRESSING
7. DINING ROOM
8. LOUNGE
9. KITCHEN
10. DOUBLE VOLUME
11. TERRACE
12. POOL
13. BAR
14. LAUNDRY
15. GYM

**GROUND FLOOR LAYOUT**
## HEAD 1818
1 : 200

客户拥有一块位于 Head Road 的陡峭地皮，希望充分利用场地，按照自己的风格和生活方式来设计房屋。他们希望拥有一栋能够稳稳耸立在场地上的房屋，屋内设有主卧室、几间客房和一些辅助性房间。

这片场地位置优越，面向海洋，背靠山峰。Head Road 的分区很有特色，其边界线两侧留有大片区域，对道路两边的斜坡上的承重也有限制。除了这些限制以外，场地旁边都是高大的建筑，对这片场地也有影响。然而，由于场地恰好有一定坡度，因此不会影响房屋的轮廓。

该项目充分利用了房屋后部的"狮头"、海角和大西洋海岸风光。这些有趣的动态风景决定着卧室的位置安排，同时还要保证空间的隐密性。客厅和卧室在纵向上有着明显的差别。最顶层是宽敞的井放式主卧室，内设更衣室和浴室，空间的前后都是壮观的风景（露台面向海洋，还能居高远望"狮头"）。

门厅处的木楼梯穿过两倍高空间，可直接通往主客厅和娱乐空间。

餐厅、休息室、厨房和早餐用餐空间采用了开放式的布局方法，与露台、花园和水池无缝相接。在这个多功能的空间中，平坦的天花板和大推拉门十分引人注目。

房屋有着清晰的现代线条。

挑战中孕育着机遇。屏风既隐藏了空间，又充当了美景中的一部分。实体结构与透明结构之间的对比因为材料的选择以及穿孔面板的引用而更加明显。设计突出了舒适感和现代、当代审美。建筑语言大胆而有力，用穿孔玻璃墙以及平面墙组成的图画式的外观布局，给空间引入了丰富的海洋风景。

场地棱角分明，因此在利用这片珍贵的场地时，给建筑也加入了一些复杂的连接点，突出了特定元素的棱角。木材的天然触感和它在空间的表现方式，给人丰富的感官享受，比如踩踏硬木楼梯板时发出的声音，以及木板散发出来的味道，奠定了整个空间的格调。这是一栋有着迷人风景的非凡住宅。更重要的是，宽阔的起居空间与户外生活空间联系紧密。这样的狭小而陡峭的场地往往不尽如人意，然而，由于室内和室外空间的相连，房屋的中心能够打破界线，成功地将空间连为一体。

房屋上下结构分明，顶楼 3 层包括客厅和阳台层，门厅和车库层，以及主卧室层。下方结构分为 2 层，包括水池、多功能客房、员工宿舍和植物栽培训。此外，还有设有通往各层的电梯。

景观是设计中的关键部分，由 Franchesca Watson 设计而成，其特色在于用茂盛植物组成园林景观。院中的铝藤架也十分有趣，随着它的缩放比例的不同，各种景观区的大小也不同。植物区让入口区域和池边露台更加舒心，由于房屋在场地上呈阶梯状，因此植物也充当了房屋横向边界处的屏障。

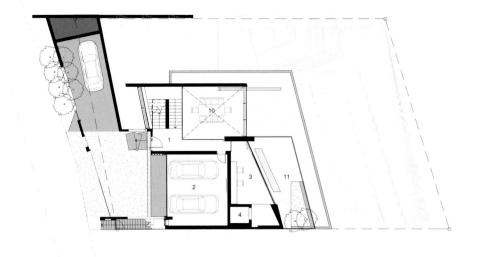

**FIRST FLOOR LAYOUT**
# HEAD 1818
1 : 200

**LEGEND**
1. MAIN ENTRANCE
2. GARAGE
3. STUDY
4. LIFT
5. BEDROOM SUITE
6. DRESSING
7. DINING ROOM
8. LOUNGE
9. KITCHEN
10. DOUBLE VOLUME
11. TERRACE
12. POOL
13. BAR
14. LAUNDRY
15. GYM

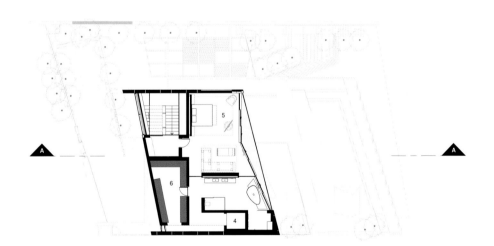

**SECOND FLOOR LAYOUT**
# HEAD 1818
1 : 200

**LEGEND**
1. MAIN ENTRANCE
2. GARAGE
3. STUDY
4. LIFT
5. BEDROOM SUITE
6. DRESSING
7. DINING ROOM
8. LOUNGE
9. KITCHEN
10. DOUBLE VOLUME
11. TERRACE
12. POOL
13. BAR
14. LAUNDRY
15. GYM

# 438 N. Faring Rd

## 法林路 438 号别墅

**Contributor**
The Agency

**Location**
Holmby Hills, Los Angeles

**Photography**
Simon Berlyn

This spectacular new estate sits behind private gates on a lush property in Holmby Hills, Bel-Air, one of the most sought-after addresses in the world. Framed on all sides by mature trees and greenery, the contemporary 3-level, glass-encased compound boasts panoramic views of the Hollywood Hills, exquisite amenities and wonderful indoor-outdoor flow throughout.

The scale and proportions of this home are breathtaking. Secluded and utterly private, the 1,486 $m^2$ home was designed and built by renowned, London-based Quinn Architects and Estate Four, with a sophisticated sense of space. The main entrance is dramatic, offering an immense motor court and impressive 6.10 m high canopy. This leads you to the luxuriously spacious living room, with its earth-toned color scheme, floor to ceiling glass, and soaring 6.10 m high ceiling, tailor-made for displaying large-scale artwork.

The home's interiors feature the highest quality materials throughout, from a marble-clad stairwell to custom recessed lighting to the American black walnut flooring. You'll enjoy an abundance of natural light and high ceilings in every room. Each space feels like a suite unto itself, and almost every room boasts its own outdoor patio or balcony with views. As you travel between rooms, dramatic vistas unfold. Transparent facades and sliding doors fuse indoor and outdoor living to exhilarating effect.

Off the 22.86 m central gallery — ideal for an art collector — the second living area boasts a 4.3 m high ceiling and direct access to the outdoor terrace and pool. An ultra-spacious kitchen features a state-of-the-art Molteni design with integrated lighting, Dada/Miele appliances and direct access to the terraces and custom-tiled infinity pool. The resort-like grounds are ideal for relaxing and entertaining, with 604 m² of terrace space in addition to the pool, landscaped gardens, outdoor kitchen, flat lower lawn and a gorgeous astro-turf tennis court with floodlights. Mature

trees provide a natural border of the property, and together with the tiered pool and gardens, it feels as if you are on a luxurious island looking out over your own private park.

Upstairs, the wide-open master suite combines a sophisticated master bedroom, huge master bath, 3 private walk-closets, powder room and separate offices. The master also boasts an extensive balcony looking out over the pool and tennis court.

The 3-level home also features 6 other elegantly appointed bedroom suites, 3 living rooms, formal dining room, 2 libraries/studies, double-height gallery, spacious fitness center, home theater, catering kitchen, elevator and wine cellar. The estate integrates a cutting-edge audio-visual system that uses an iPad interface, everything automated and with maximum security. This is truly an estate for the ages.

这栋华丽的全新私人豪宅坐落在霍姆比山最受欢迎的高档住宅区贝沙湾。这栋 3 层楼被茂密的植被拥簇着，拥有整个好莱坞山庄的壮丽风光，室内和室外景致都同样的令人心旷神怡。

房屋是由伦敦著名的 Quinn Architects 和 Estate Four 两个机构合力打造而成，基于其丰富的空间设计经验，他们将这个占地 1 486 m² 的建筑打造成了隐密的私人空间，不论是建筑的规模，还是建筑的比例，都令人赞叹不已。房屋的架构和区间比例在设计师的精心规划下，让人心醉神迷。正门场地开阔，由大停车场和 6.10 m 高的华盖组成，穿过正门就来到了富丽堂皇的会客厅，从地板到天花板都采用了怡人的大地色系，6.10 m 高的天花板则是为展示大型艺术品专门设计。

房屋的内饰全部采用高档原材料，大理石楼梯井和深色美洲胡桃木地板，处处彰显出典雅的气息。在每一个房间，你都能感受到完美的自然采光和空间的宽阔感。每个空间都像套间，搭配协调，甚至几乎每个空间都带有露台和观景阳台。当你在空间与空间中穿梭，壮丽的远景便在你眼前慢慢呈现。透明的墙壁和滑动门将室内和室外空间连为一体，产生惊人的效果。

22.86 m 长的中央走廊是艺术收藏家的最爱；次客厅的天花板高 4.3 m，能直接通往户外露台和水池。宽敞的厨房内设有德易家（Molteni）的先进设计、集成照明设施、大达（Dada）和美诺（Miele）家用电器。从厨房可以直接通往露台和无边水池。该住宅就像度假屋一样，是放松娱乐的理想场所，除了水池、花园、户外厨房、平坦的低草坪和设有强光灯的高档人造草皮网球场外，还有 604 m² 的露台场地。茂盛的树木是建筑的天然屏障，配上层叠式水池和花园，给人以一种居于奢华之岛，远观属于自己的公园的感觉。

楼上设有开敞式的主套房，并设有大床、大浴室、化妆间、独立办公室和 3 个私人步入式衣柜。站在宽敞的阳台上能看到水池和网球场。

这座 3 层住宅带有 6 个雅致的套间卧室，3 个起居室，3 个衣帽间还有 2 个独立的书房办公室。双层高的画廊，宽敞明亮的健身房，欧式厨房、室内电梯、酒窖、影音娱乐室、私人泳池、私人网球场，各类设施一应俱全。先进的自动化视听设备设有 iPad 接口，安全性极高，是老人的理想退居场所。

# Winelands 190

190 号酒庄

**Architecture & Interior Architecture**
Antoni Associates

**Photography**
Adam Letch

**Project Team**
Mark Rielly, Sarika Jacobs, Jon Case, Clive Schulze

**Key Furniture Supplier**
OKHA Interiors

**Location**
Stellenbosch, Cape Town

The clients' brief was to ensure that when they occupied the house on their own that it was not too big and empty, however at the same time it also had to allow for the whole family and grandchildren should they decide to stay over. The decision was made to keep all the entertainment spaces as well as the master bedroom on the ground floor, with three additional guest suites on the first level. As the De Zalze design regulations only permitted a single story, the guests' suites were accommodated in the roof attics.

Two important criteria for the clients were that the project had to include a central courtyard and that the main living areas were to have double volume. At the front of the plot the concept of a traditional "Cape Dutch Langhuis" with gabled ends was

conceptualized. This part of the house was designed to contain the formal lounge & dining, informal lounge & braai room as well as the master bedroom and en-suite, giving all these areas direct access to the front pool terrace with spectacular views to the surrounding vineyards and mountains. In the "Langhuis" open trusses were used to give all the rooms a larger volume. To the rear of the "Langhuis", the two side wings and a linking lobby create a sheltered central courtyard which also visually links the kitchen to the family TV lounge. Identical feature stairs link to the guest bedroom suites located in the roof lofts of these back wings.

The owners of this home love to entertain and wanted the house to reflect their lifestyle. Key features of this include the bespoke wine cellar. Here the design team created a spectacular glass wine wall. The oak timber cabinetry encases frameless glass shelving which can house more than 400 bottles which are perfectly temperature controlled with concealed refrigeration. The wine wall also functions as a visual screen between the formal and informal family spaces. A dramatic double volume stone clad fireplace in the formal lounge is mirrored by a built-in braai in the informal lounge. These areas of the house flow out onto the outdoor entertainment deck and infinity pool. Linked to the terrace is a sunken outdoor boma (a typical South African outdoor enclosure). Here casual seating is arranged around an open fire.

For the interior architecture the design approach by Mark Rielly and Jon Case was to focus on the use of natural organic materials such as timber and stone. Limed oak flooring paired with honey coloured stone walls contrast with black charcoals and chalky white finishes. These tactile materials add a sense of homeliness and warmth to the contemporary architecture. A number of elemental forces are captured in the use of water features and fireplaces.

Werf walls and a pergola covered walkway lead to the front entrance which opens into the courtyard lobby overlooking the reflective pond and greenery. Glass pocket doors create separate entrances and lead into the side wings. Focal features of the entrances are the floating sculpture ledges. Here Angus Taylor sculptures are reflected in the fractured mirror wall

cladding. The same fractured pattern is again used in the smokey mirror cladding of the stairways. Limed oak cantilevered treads, timber cladding with white stone ledges and frameless glass balustrading add to the transparency and airiness of these spaces. Clear blown glass globes by David Reade are suspended and reflected in the double volume stairwells adding interest.

A combination of bold and discreet lighting was used to create a "wow" factor and the layering of lighting set various moods. Subtle lighting has been incorporated in all recesses and feature bulkheads to give a warm glow to peripheral edges. Concealed lighting has also been used to highlight and accentuate the organic natural finishes. In the dining room a customized crystal chandelier by Martin Doller is suspended from the ceiling rafters. Other feature lighting includes a custom designed "ring" light by AA Interiors, over the informal braai room and a signature Willow Lamp in the main bedroom.

The interior furniture and décor were designed by Mark Rielly and Sarika Jacobs of AA Interiors. The furniture is modern and complementary to the experience of the home. Tactile finishes including timber, textured leathers and raw linens add a sophisticated sense of understated luxury. The clients' love for colour has been introduced with injections of bold prints and vibrant fabrics. Bespoke furniture from OKHA Interiors is featured throughout.

客户希望当他们住进房屋时，不会感觉房屋太大、太空，同时，空间能够住下整个家庭和有时会留下来住的孙子和孙女。设计方案是将娱乐空间和主卧室设在一楼，而其他3间客人套房设在二楼。由于 De Zalze 的设计规定只能建一层，因此客人套房设在了阁楼上。

按照客户的要求，项目中应当包括一个中央庭院和两倍高的主起居空间。起初，设计师构思了一个传统的三角墙"荷兰风开普敦建筑结构"，其中包括正式客厅和餐厅、非正式客厅和烧烤房、主卧室和套房，而且所有空间都朝向水池和露台，远观壮丽的葡萄园和山峰风光。该建筑结构内的桁架使所有空间更有空间感。该结构后部的两翼和连接两翼的大厅组成了遮阳中央庭院。在中央庭院中能看到厨房和电视休息室。功能相同的楼梯可以直接通往后翼阁楼上的客房。

房主热爱闲适的生活，因此希望房子能体现这种生活方式。最能体现这一点的莫过于量身定做的酒窖了。设计师为酒窖设计了一道壮观的玻璃酒墙。无框玻璃酒架被橡木橱柜包围起来，至少可以容纳400瓶酒，而隐藏式的制冷系统则将温度控制到最佳状态。除此之外，玻璃酒墙还充当了正式家庭空间和非正式家庭空间之间的隔墙。正式客厅中的壁炉上镶着两倍高的大石块，与非正式客厅中的嵌入式烤炉形成对比。这些空间都与户外休闲露台和无边水池连通着。与露台相连的是下沉式户外圈地（典型的南非户外围场），此处的户外火炉被各种休闲座椅包围着，颇为惬意。

住宅的室内装潢是由 Mark Rielly 和 Jon Case 设计的。他们注重天然有机材料的使用，例如木材和石材。在这个空间中，用石灰处理过的橡木地板与蜂蜜色石墙相呼应，又与黑色的炭石、白垩色的家具形成对比。这些颇有触感的材料使这栋当代建筑更加温馨，而水景和壁炉则展示出了自然元素的力量。

"沃夫墙"与藤架围绕着通往门厅的走道,门厅敞向庭院,在此处能俯瞰清澈的水池和茂盛的绿植。玻璃推拉门能将门厅封闭起来,而推开这道门,可以直接通往房屋的侧翼。门厅中最引人注目的是悬浮的雕塑板。而 Angus Taylor 的雕塑作品则是被倒映在了碎裂的镜墙中。楼梯处的烟灰镜墙同样也是碎裂的。悬臂式的楼梯是由用石灰处理过的橡木踏板、白色石块底座和无框玻璃护墙组成,使得这片空间十分通透、空气畅通。由 David Reade 设计的吹制透明玻璃球悬挂在楼梯的上方,给这个两倍高的空间增加了趣味性和空间感。

醒目的灯具与素雅的灯具相结合,营造出惊人的灯光效果,层层灯光照射下来,形成不同的空间氛围。微妙的灯光覆盖了整个内部空间,而别致的灯头照亮了房屋的

外围空间，营造出一种亲切的氛围。而隐藏式的灯具又有着不同的角色，其灯光突出了空间内各种物品的天然肌理。由 Martin Doller 专门设计的水晶吊灯从餐厅屋顶橡梁垂下来，美不胜收。其他的灯具包括非正式烤房中由 AA Interiors 设计的环形灯，以及主卧室中的柳木灯。

室内的家具和饰品都是由 AA Interiors 的设计师 Mark Rielly 和 Sarika Jacobs 设计的。现代家具契合了这栋房屋的现代格调。各种物品都是由木材、皮革和粗亚麻布制成，不仅触摸起来十分舒服，而且给空间增添了一种低调的奢华感。印花布的艳丽色彩都是房主喜欢的色彩。除此之外，OKHA Interiors 设计的家具也被应用到了空间的各个角落。

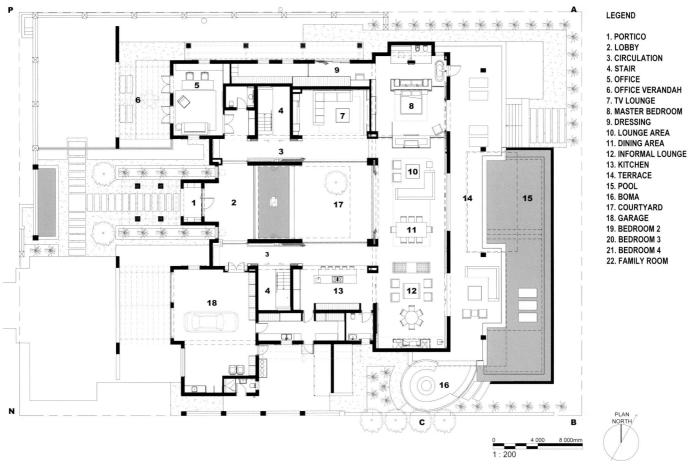
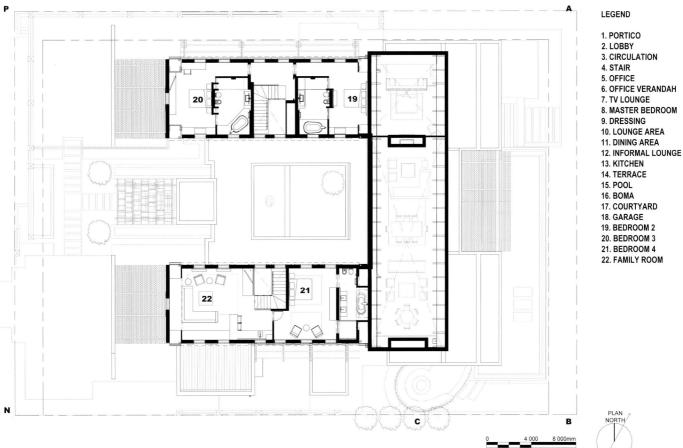

# Expressing Views

## 观景别墅

**Designer & Builder**
Urbane Projects Pty Ltd.

**Location**
Perth, Western Australia

**Photography**
Joel Barbitta of D-Max Photography

Contemporary, timeless and spreading over three levels with a sparkling infinity edge pool to its elevated corner block, this Apple cross abode commands attention from first glimpse and allows its family of four to live the lifestyle they had always dreamed about.

The client shad lived for years in the older house on the block. When the chance arose to buy the house next door they took advantage of it to knock down both residences and build their dream home with Urbane. Now they have their "forever" home with a level of luxury finishes and a custom design that does justice to its market riverside location and makes the most of the site's panoramic river views.

A 3 m tall ebony-stained jarrah door opens onto the entrance foyer. A wall of

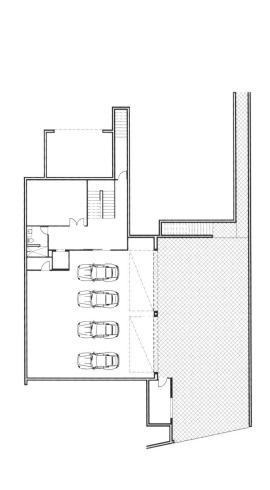
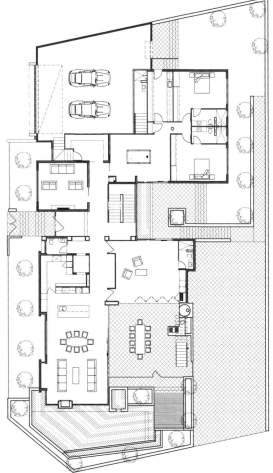

windows overlooks a reflection pond, mature frangipani tree and sculpture set against a backdrop of a glass face rendered wall. In the entry is one of the home's most spectacular features — the staircase. A feat of engineering, the staircase runs through all three stories and features a suspended steel structure with oak treads stained a rich Japan black that complement the honed limestone floors.

The clients wanted the design to have longevity and to cater for their teenage daughter and son for years to come. 5 bedrooms and 7 bathrooms encompass separate zones for the teenage children, as well as guest bedrooms for friends to stay over. A games room as well as a home theatre offer spaces for the children to entertain their friends, while parents have an upstairs bar and lounge, as well as a fully enclosable alfresco lounge and dining with its own bar, all set against river views. The hotel-like parents' retreat has a wraparound balcony and views from the bed and ensuite that stretch across the neighbourhood to the river and city beyond.

The open-plan kitchen, lounge and dining can also be opened up to the outdoors for effortless indoor-outdoor living and entertaining. The landscaped gardens include an infinity edge pool tiled in iridescent Bisazza mosaics that visually link the gardens to the sparkling Swan River beyond.

The clients said they were thrilled with Steve's work and loved the Urbane approach, encompassing everything from conception to the design of the custom-built cabinetry and selection of furniture. "Everything Steve picked, we loved." they said.

这栋当代风格的 3 层楼建筑可谓经得起时间的考验，波光粼粼的无边水池抬高了角落区域，建筑呈交叉苹果状，十分引人注目，实现了一家 4 口对理想生活方式了追求。客户在该区的老房子中住了多年，恰好有机会购买隔壁的房子，便欲拆掉原有的两栋房子，与 Urbane 合作打造梦想家园。现在，他们拥有了一个装饰豪华、设计独特的"永恒"家园。这栋别墅与滨河区域的高档氛围相一致，该地大部分的河景也尽收眼底。

推开 3m 高的黑色红木大门，步入大厅，便能透过玻璃幕墙看到屋外清澈的水池，茂盛的赤素馨花树和雕塑都以玻璃墙为背景。入口处的悬臂式楼梯是房屋的亮点之一，这一工程上的壮举贯穿了房屋的 3 层空间，其钢架和橡木踏板都被漆成饱满的日本黑色，与光滑的石灰岩地板相呼应。

客户希望房屋能经得起时间的考验，并能满足儿女的需求。房屋内设 5 间卧室、7 间浴室和客房，其中还包括专为儿女青少年时期打造的私人空间，甚至还有为孩子及其朋友们打造的游戏室和家庭影院，而家长则可以在楼上的酒吧和休息室、可封闭的露天休息空间和带酒吧的餐厅中一边生活，一边欣赏美丽的河流风光。这栋酒店式的家庭度假屋被阳台环绕着，卧室和套间中都能远眺欣赏临近的房屋、河流以及城市的风景。

开放式的厨房、客厅和餐厅敞向户外空间，轻松地将室内和室外生活与娱乐融为一体。景观花园中设有无边水池，池中铺满了彩色碧莎马赛克瓷砖，放眼望去，似乎与远处那波光粼粼的天鹅河相接。

客户对史蒂夫的设计十分满意，Urbane 所采用的手法，包括从设计到定制橱柜，再到挑选家具的一系列过程得到了客户的肯定。"我们喜欢史蒂夫所挑选的一切。"客户说道。

# Laurel Way

## 月桂街别墅

**Architects**
Marc Whipple AIA

**Project Manager**
Andrew Takabayashi

**Interior Designer**
Michael Palumbo

**Location**
Beverly Hills, California

**Area**
931 m²

**Photographers**
William MacCollum, Art Gray Photography

One aesthetic idea driving the creation of Laurel Way was that each room or space should be a jewel box, an individually conceived, precisely functional and dramatic sensory experience with its own depth of architecture.

Central to the composition are many of Marc Whipple's signature elements, one being the use of texture; smooth next to rough stone, rich wooden panels against glass, and glass reflecting water. The immediate experience upon entering the house is its inherent weightlessness — the sense that the walls appear to float as panels and you are always connected to the outdoors. This is achieved with adherence to precise symmetry of beams, support panels, tiles, and sightlines, and also that walls do not meet the ceilings — a 1.3 cm gap is left that helps achieve the effect.

These elements play up the horizontals and verticals of the house while movement and curves come from the 3 tiers of greenery and 2 water channels that surround the house giving it the look of an island floating against the blue California sky. The moat-like water surround is more than a successful artistic inspiration; it adds the feeling of a protective boundary without obstructing the views in any way. It also provided an innovative water feature visible from the interior while adding a highly dramatic dynamic to the entire design.

The front entry steps lead to a 4.3 m wood pivot door flanked entirely by glass, and then into the main floor foyer. To the left, a section of glass flooring reveals a wine room below with storage for 1,000 bottles, and cantilevered wenge wood stairs float upward to the bedrooms.

The living and dining areas are a study in chocolate and creamy whites carried through to the exterior surfaces achieved with Texston's Lime based plaster, offset by rough split-faced stone and dark wenge wood. Lift and Slide German made Schuco windows and doors are state of the art offering dependable operation and drainage as well as thermal efficiency summer and winter. Glossy kitchen cabinets were custom designed and imported from Italy.

"Zero edge" and " floating" themes are echoed in the smallest details; kitchen cooktop venting is flush to ceiling.

With no use of molding all lines are visible, every element must be perfectly square and aligned. Minotii, Maxalto and B&B Italia furniture was selected or custom made for each living space. The main powder room's motorized sliding glass door opens up to a vanity and white glass rectangular column — the sink. A wall of small, mirrored black tiles, reflect a single chrome vertical water pipe suspended over custom made sink.

　　月桂街别墅的设计是受一种美学理念的启发：每个空间都应该是一个珠宝盒，其构思和设计都是依据各个空间的用途，并建筑的深度来给人以强烈的感官体验。

　　项目的核心结构中包含了许多马克·惠普尔（Marc Whipple）的主打元素：质感对比手法——光滑与粗糙的石头、丰富的木板与玻璃墙、透明的玻璃与清澈的水。进入空间，人便感觉房屋有种失重感：墙壁和壁板都悬浮在空中，而且室内与室外空间总是相互连接着。这种效果的实现得益于横梁、支撑板和瓷砖之间对称的精确性，墙壁和天花板之间留有 1.3 cm 的高度，视线可以毫无阻碍地从中穿过。

　　当 3 层绿植和 2 道水景在整栋房屋内运动形成一道道曲线时，这些元素便组成了房屋的横向和纵向结构。房屋就像是漂浮在加利福利亚蔚蓝天空下的岛屿。护城河般的水景环绕着房屋，不仅成功地展现了艺术的灵感，还在不遮挡任何角度的风景的条件下，增加了一条防护线。同时，这一创意水景还充当了室内空间的风景，给整个设计增添了强烈的动感和活力。

　　正门前的楼梯直接将人引向 4.3 m 高的木转轴门，门的两侧都是玻璃墙，接下来便是主层门厅。透过门厅左侧的玻璃地板，就能看到楼下的可以储存 1 000 瓶酒的酒窖，悬臂式的崖豆木楼梯可以直接通往楼上的卧室。

　　客厅和餐厅空间采用了巧克力色和奶油白色，此色调一直延续到 Texston 石灰外墙，又被粗糙的裂纹石和深色崖豆木中和了。最新的舒克（Schuco）上翻式和滑动式门窗对调节冬夏季室内的湿度和热度都十分有效。

　　"零界线"和"漂浮"这两个主题与微小的细节相呼应，厨房的灶台上方的排气设施嵌入天花板中。

　　所有的线条都未加修饰，清晰可见；各元素排成一列，相互对齐。设计师还为每个生活空间精心挑选或定制了 Minotii、Maxalto 和意大利 B&B 家具。主盥洗室设有自动滑动玻璃门，敞向洗手间和白色玻璃方柱盥洗盆。小黑色镜面砖镶满整面墙壁，倒映出垂直悬在盥洗盆上方的铬水管。

# River House

河畔别墅

**Architects**
MCK Architects

**Landscape Architect**
The Potager Garden

**Builder**
Alvaro Brothers

**Planning Consultant**
Mersonn

**Area**
530 m²

**Photography**
Steve Back

**Interior Decoration**
Scala and Romano Interiors

**Structural Engineer**
Simpson Design Associates +
Luke Tsougranis

The River House is essentially a home that integrates an informed assembly of clean lines, warm textured materials and daring structure on a difficult site.

The inhabitants now have a bespoke living vessel to live, work, and play within, and it has arguably created spaces that allow the members of the family to better congregate. The design allows the family and visiting guests to comfortably integrate or disperse for privacy as required. The house does not appear as a very big house through careful manipulation of the massing. The visual program is split into distinct objects which further breakdown the overall massing. Internally it has a series of clever buffers that separate distinct zones.

**Ground Floor Plan**

01 Entry
02 Hall
03 WC
04 Living
05 Kitchen
06 Dining
07 Terrace
08 Deck Area
09 BBQ
10 Study
11 Laundry
12 Guest Bathroom
13 Guest Bedroom
14 Bin Store
15 Garage
16 Lower Garden
17 Daybed below
18 Lower Pool Yard
19 Pool below

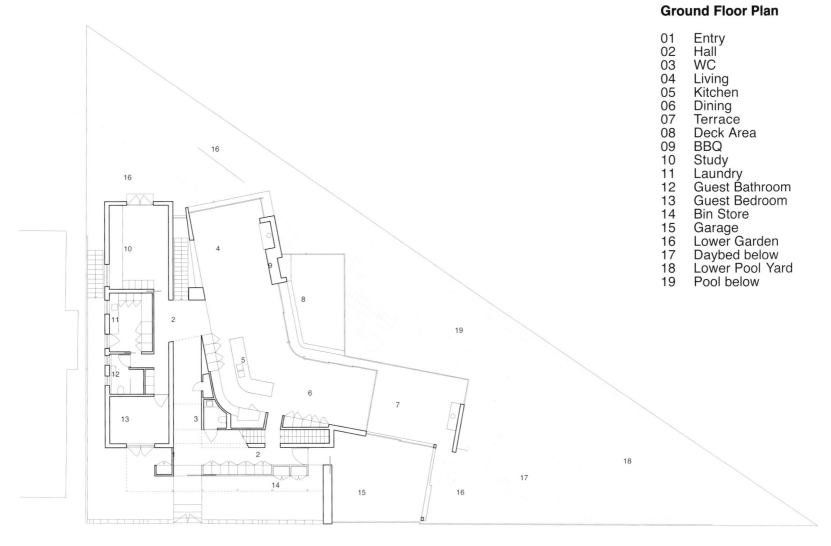

**river house**

**Ground Floor Plan**

0  1          5          10

The house takes on a schizophrenic nature by virtue of it's conservative, and private streetscape versus the more wild and open, cantilevered living zone. The streetscape adopts textures, colours and materials from the locality, specifically shingle, sandstone, and painted brick.

The functional performance of the design is wholly based on the clients brief busy lifestyles. The functions are split to 4 levels through the site being from top to bottom, parent's retreat, living/guest/work level, children's sleeping playing level and the pool/backyard level. The latter being newly constructed some 4.5 m above the existing yard, to eliminate an existing inaccessible space and better connect the "fun" parts of the design to the house proper.

The River House proved itself to be one of our greatest collaborative efforts to date. Between an incredibly enthusiastic and intuitive client, an innovative structural engineer, a master craftsman in a builder and ourselves, the design took shape and pushed everyone's imagination to their limits. At the end of the day this project would not have been achievable without the "integration" of such "allied disciplines".

There was a big push to retain a lot of the existing thick basement sandstone walls as they represented the original fabric of the building and a high quality of building stock. The decision to cantilever the main living level naturally increased construction costs however it allowed the design to be maximised on a smaller awkwardly shaped site, and effectively made the most of a smaller site. The abundance of controlled natural light also diminishes the requirement to use artificial lighting when the sun is shining.

Sustainable principles in design include retention of quality building stock like the 600 mm thick lower basement sandstone walls, implementation of a 20,000 L rainwater tank, retractable solar blinds to harness natural light as required, and means to enhancing cross ventilation. Laser cut security screens allow the house to breathe in the warmer months whilst providing adequate security, thus reducing the need to use air-conditioning.

The main living levels are constructed using thick concrete slabs creating excellent thermal mass qualities. General orientation of the plan enables best solar access.

**First Floor Plan**

20 Main Bedroom
21 Void
22 Ensuite
23 Deck Area
24 Roof
25 Deck below

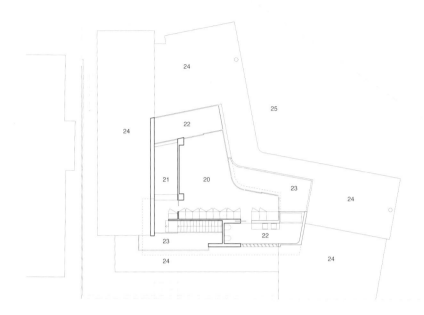

river house          **First Floor Plan**     0 1    5    10

　　清晰的线条、质感柔和的材料和大胆的结构构成了这栋河畔别墅。虽然它所处的场地高低不平，但它却是一处理想的家园。

　　在这些创造的空间中，客户及其家人可以生活、工作、嬉戏，更好地相聚于其中。房屋设计满足了这个家庭和来访的客人要求，不管是相聚的热闹，还是独处的宁静。建筑经过精心雕琢显得小巧玲珑。人的视线先是被醒目的物品吸引过去，随后，又落到整个结构上。屋内一系列的设置、巧妙的过渡区，分隔了各个特色空间。

　　建筑的保守风格，配上独有的街景，以及更自然、更开放的悬臂式生活空间，使房屋具有多种特征。街道景观采用了当地独有的木瓦、沙岩和油漆砖，这些材料有着特别的质感、色调。

　　功能空间的设计完全是基于客户忙碌的生活方式。功能区从上至下分为 4 层：父母休息区、起居客厅和工作区、孩子睡觉玩耍区和水池后院区。新建的 4.5 m 水池后院取代了原有的花园，打开了原来无法进入的区域，更好地连接了房屋设计中的娱乐空间。

　　作为设计师倾尽心血的作品之一，该别墅是热情敏锐的客户、有创意的结构工程师、建筑公司的工艺大师与设计团队密切协作的成果，是将所有人的想象力推向极致而成形的设计作品。如是没有各方的合作，项目是无法完成的。

　　大部分厚沙岩墙基也经过了一番努力才得以形成，展现了建筑的原本构造和较高的建筑质量。将主起居层悬起来的决定自然是增加了建设成本，但同时也了增大了对这形状怪异的小场地的设计可能性，有效地利用了周围的小场地。可控的自然光十分充足，当阳光灿烂时，室内对灯光的需求便大大减少。

　　设计遵循了可持续原则，保留了有用的建筑储备区，如 600 mm 厚的低地下室沙岩墙；还设有 20 000 L 的蓄水池和折叠遮阳百叶窗；此外，还采用了一些增强对流通风的方法，如激光切割的防盗纱窗在暖和的月份给屋内带来了新鲜空气，减少了空间对空调的需求，同时也有较强的防盗功能。

　　主起居层是用厚混凝土楼板建成，具有良好的储热能力。房屋规划的总体朝向合理，室内空间光照充足。

# Fieldview

## Fieldview 别墅

**Architects**
Blaze Makoid Architecture

**Area**
372 m²

Located on a flat, one acre flag lot with neighbors close to the front and side yards, this 10,000 m² house is configured of 3 primary volumes arranged in an "C" that frame the expansive, southern view of an adjacent, agricultural reserve. This view serves as a backdrop to an interwoven composition of interior and exterior spaces.

Entry, through a glass void in the northern side of the house, is approached by a raised, stone walk, under an exaggerated uplighted canopy. The entry foyer, at the terminus of the outdoor pool, separates public space to the left and the private, 2-story bedroom wing to the right. An open floor plan contains living room, dining room and kitchen stretches along the length of the central outdoor patio. Large expanses of south facing glass help to dissolve interior/exterior relationships while a more

selective glazing strategy locates individual windows in the predominately solid north, east and west walls that create privacy while modulating temperature.

The arrangement, assisted by a series of pushed and pulled planes maximizes the ability to modulate the various sunlight requirements while creating more intimate indoor and outdoor functions that serve various functions as activities migrate throughout the day — swimming, breakfast, sunshine, lounge, sleeping.

该别墅坐落于 10 000 m² 的平地上，屋前和侧院都有邻居。建筑占地 372 m²，分为 3 个主要空间，呈 "C" 字形展开，饱览近处广阔的农业保护区风景，它们也成为了相互交织的室内空间与室外空间的背景。

踏上高台石道，穿过熠熠生辉的大型华盖，跨进房屋北侧的入口，便到了以玻璃幕墙覆盖的宽敞空间。露天水池尾端的门厅将左侧的公用空间和右侧 2 层私人卧室空间分隔开来。开放式的空间布局中包括客厅、餐厅和厨房，都沿着中央的露台延伸开来。朝南的玻璃幕墙使室内与室外空间融为一体，而北面、东面和西面的实体墙的窗户更为独特，玻璃窗的种类也更多。这些窗户在调节室内温度的同时还能保证隐私。

空间布置中还加入了一系列的推拉门，从而能够更好根据需求调节空间中的光线。同时也使室内与室外空间联系更紧密。居住者在进行游泳、享用早餐、日光浴、休息和睡觉等各种活动时，能轻松在空间与空间之间移动。

/ 163

# Wind House

风之墅

**Architects**
OPENSPACE DESIGN Co., Ltd.

**Location**
Noble Residence, Bangkok, Thailand

**Client**
Janphim Sukumaratat

**Gross Floor Area**
675 m²

**Scope of Works**
Architecture and Interior Design

"Wind House" was created as "Resort Space" which was the owner's preference style regarding the site conditions. As it was located on the edge of the housing estate project's boundary, it gained the view of big natural green area beside and certainly, the atmosphere of tranquility.

The house planning started from the idea — "How to live comfortably with nature?" Therefore, the building orientation and the space of the house should allow the wind to flow through and allow natural light to shine in without too much heat. At the same time, the users inside could be able to see nice garden view outside as well.

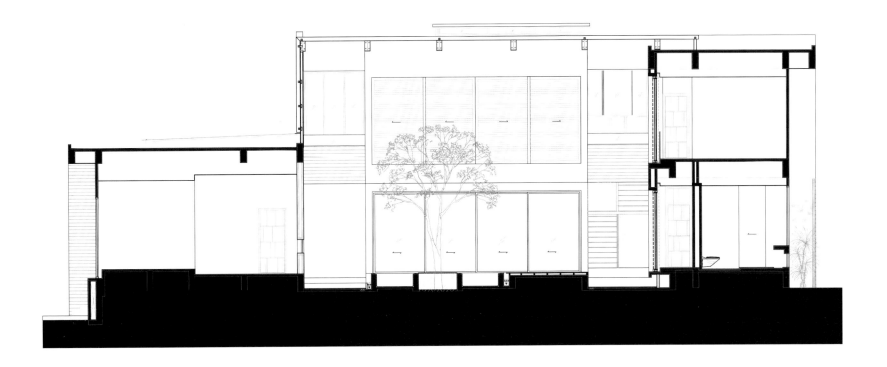

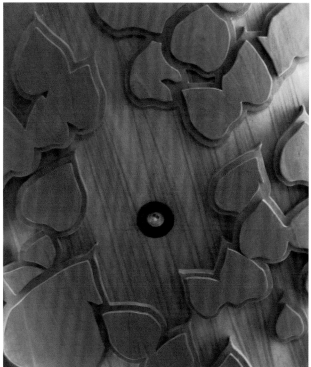

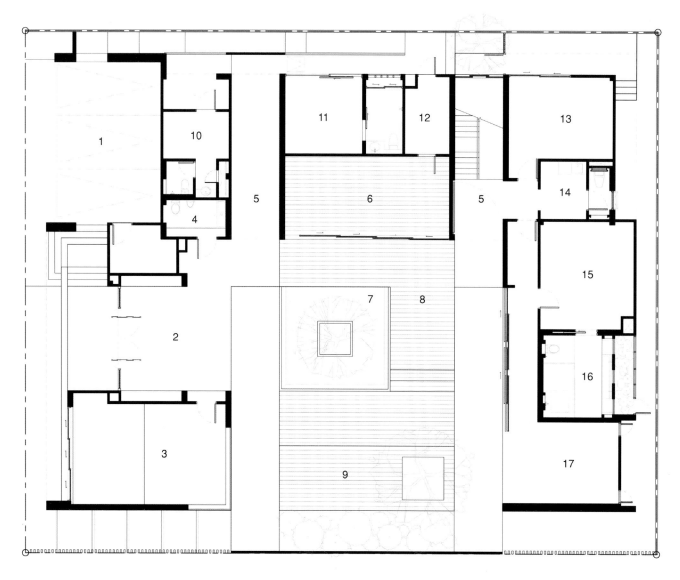

1st FLOOR PLAN

1 : CARPARK
2 : FOYER
3 : LIVING ROOM
4 : POWDER ROOM
5 : THE CORE
6 : DINING ROOM
7 : POND
8 : OUTDOOR LIVING
9 : PAVILION
10 : MAID ROOM
11 : GUEST BEDROOM
12 : THAI KITCHEN
13 : BUDDHA ROOM
14 : PANTRY
15 : WALK-IN CLOSET
16 : MASTER BATHROOM
17 : MASTER BEDROOM

The house was designed in "C" shape providing big courtyard on the right side, close to big green area beside, where every function from 3 sides could really share this pleasant courtyard together. The building itself could provide privacy to the users as neighbors would not be able to look into the center of the house. The technique to draw the wind flowing through the house perfectly was to provide some big voids of the building mass aligned to the courtyard which were also used as circulation core, stair and relaxation corner. Moreover, even some details such as doors, fences, sunlight screen patterns, etc. were meticulously designed to utilize the wind more efficiently for ventilation purpose.

One of the most significant design strategies was to create "Seamless Boundary" between building and nature, indoor and outdoor. All of the common area as well as circulation were treated as "Semi-Outdoor" space, under the roof but without walls, connecting to the courtyard harmoniously. In addition, some enclosed functions were still optional to get fresh air sometimes by sliding full-height partitions to the sides. These would enable the house space to look wider, more airy and definitely, to welcome the delightful wind to be "Wind House".

根据业主喜好的风格以及场地特征，该住宅被设计成像度假别墅一样的。由于住宅建于地产项目场址的边界上，因此，它坐拥大片天然绿地风景以及宁静的环境。

房屋规划的出发点是解决"如何在大自然间舒适生活"的问题。因此，建筑的朝向和空间布局应允许自然风易进入室内，且室内有充足的自然光线而不会带来太多热量。同时，居住者还能观赏户外花园美丽的风景。

房屋呈"C"形展开，在右侧形成宽阔的庭院，与大片绿地很近。三侧的功能空间能够共用庭院。由于周围邻居无法看到房屋的中心区，从而充分保证了房屋的隐密性。朝向庭院的建筑部分设计了许多大空隙，它们充当了核心流通区、楼梯或休闲一角，也能将自然风引入室内。此外，房屋的细节设计非常严谨，比如门、栅栏和遮阳屏等的设计都更有效达到了自然通风的目的。

住宅最重要的设计策略是打造建筑与自然、室内与室外间的"无缝边界"。所有的公共空间和流通空间都被设计成"半露天式空间"，有屋顶却无墙壁，与庭院无缝衔接。此外，一些封闭的功能区两侧是满高的滑动隔离墙，推开隔墙时不仅可将新鲜空气引入风之墅，还能让空间显得更加宽敞。

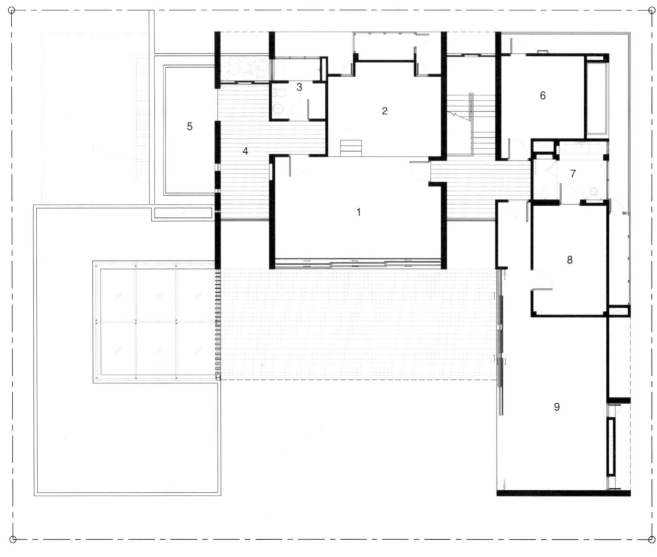

2nd FLOOR PLAN

1 : HOME THEATER
2 : WORKING ROOM
3 : BATHROOM
4 : RELAX CORNER
5 : TERRACE
6 : STORAGE ROOM
7 : BATHROOM
8 : WALK-IN CLOSET
9 : BEDROOM

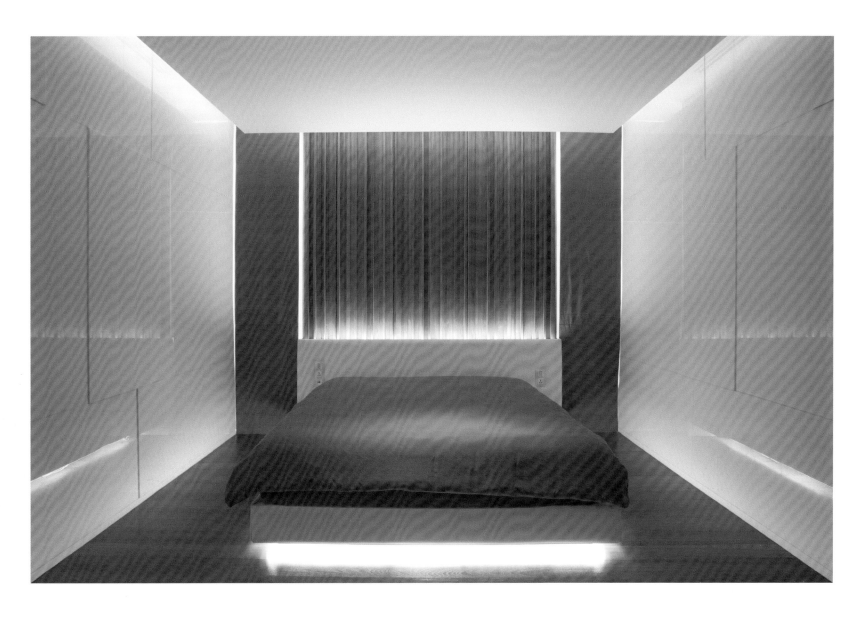

# Cadence Residence

卡当斯别墅

**Architects**
Keith Baker, Keith Baker Design Inc.

**Builder**
Taran Williams, TS Williams Construction

**Interior Designer**
Ashley Campbell, the Interior Design Group

**Photography**
Mia Dominguez, Artez Photography

Completed in the Spring of 2014 "Cadence" was designed as a series of pavilions, nestled along the sunny shoreline of Lantzville on Vancouver Island British Columbia. This stunning home was an award-winner even while it was on the "drawing board". In 2012 it won a Gold Award for "Best Home Design Concept" at the annual Construction Achievements & Renovations of Excellence Awards. And now that it is built, it is a finalist in 7 prestigious categories in the "CARE Awards" including Best Custom House, Best Contemporary Kitchen, Best Master Suite, Best Interior, Best Millwork, People's Choice and Project of the Year. Finally, "Cadence" won 'Best Custom Home over $1M" at the Vancouver Island CARE Awards. It is also the finalist in four categories in the (provincial) Georgie Awards at the end of February, as well as it has won "Best Custom Home (in Canada) over $1M" at the National SAM Awards.

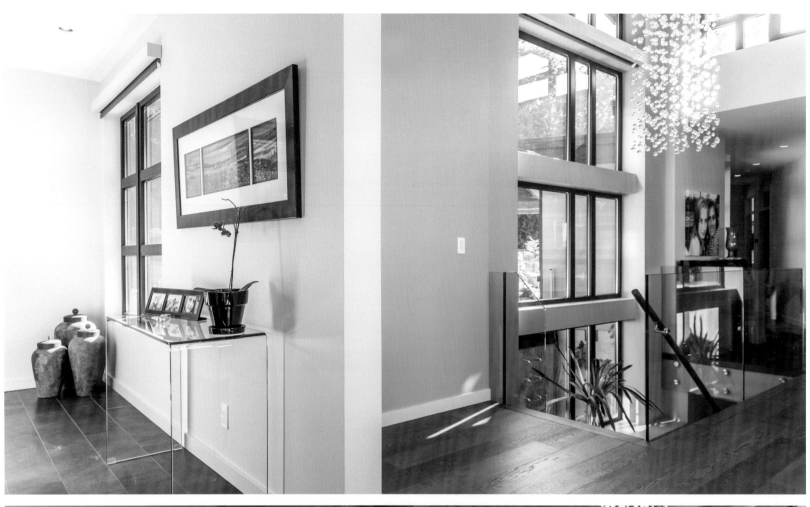

/ 177

"The concept is such that the spaces should read as very open and create a flow without being too grand. The use of scale was very important by keeping the relative forms of a human size, which gives a very comfortable natural and relaxed feeling," said Keith Baker of Keith Baker Design Inc., the designer of the home. The radiused roof undulations are a subtle reference to the waves and the ocean environment. The light filled home feels very fitting in it's location resting comfortably along the sandy shore. The materials reflect a rooted modernism including extensive Douglas Fir clustered columns and grain-matched radius beams, Western Red Cedar siding rendered in both horizontal 2.54 cm x 10.16 cm T&G and the familiar 'cottage' texture of shingles. Concrete is used as a complement and underscore to ground the composition.

It's open plan is unusual in that it is not one large great room, but rather a series of open spaces interconnected and articulated between the 2 main pavilions of the kitchen and the living room. An 5.5 m wide 3 panel sliding glass patio door opens the dining room to the entertainment sized lounging patio, the ocean and the evening sunsets. As well the outdoor kitchen is well equipped with a wood burning fireplace, a gas fireplace for gathering around, a pizza oven, concrete countertops and BBQ.

One of the many standout features is the post and beam covered breezeway that offers a graceful transition from the triple garage to the mudroom entry and separate guest suite entry.

The Porte Coachiere lends a protective elegance for guests arriving at the front door. The home also features a master bedroom suite that at once is cocooning and cosy as well is expansive and connected to nature with stunning beach and oceans views. All in all an inviting open and intimate place to call home.

卡当斯别墅于 2014 年春正式完工。别墅坐落在不列颠哥伦比亚温哥华岛兰茨维尔海岸上，在这个精妙的楼亭群设计尚未完工时，便已获得一些奖项。2012 年，别墅在"年度建设成就和优秀装修"比赛中获得了"最佳家居设计理念"金奖。如今，别墅已经建成，并获得了"CARE Awards"奖 7 项提名，这些奖项包括"最佳定制住宅""最佳现代厨房""最佳卧室套房""最佳室内设计""最佳木工""人民选择奖"以及"年度最佳项目"。最终，该别墅还在温哥华岛"CARE Awards"中被评为"价值过百万的最佳定制住宅"。2014 年二月底，该项目还获得了省级"乔治奖"4 项提名。此外，该项目在国家级"山姆奖"中荣获了"价值过百万最佳定制住宅（加拿大）"奖。

"项目的设计理念旨在创造非常开放的空间，有着流畅的空间过渡，而占用的空间又不会太大。规模手法的应用十分重要，它采用了与人体外形相当的形式，给人以舒适、自然、轻松的感觉，"别墅的设计师 Keith Baker（Keith Baker 设计公司）说道。弧形的屋顶设计巧妙，与起伏的波浪和海洋的环境相似。房屋建在沙滩边，明朗的室内空间与海岸环境十分谐调。材料的选用，包括大量的花旗松柱子、纹理半径匹配的横梁、2.54 cm× 10.16 cm 的红雪松护墙板以及熟悉的木屋采用的木瓦，体现了现代风格。而混凝土只是一种补充，仅仅充当了房屋的基座，被埋藏在地下。

开放式布局并不意味着一个单一的大空间，而是一系列相连通的开敞空间，就如厨房和客厅所在的两栋楼亭一样，衔接在一起；有 3 扇 5.5 m 宽的露台滑动玻璃门可直接通往客厅，进而再通往休息露台，便能观赏到海洋和傍晚的日落风光。户外

厨房设施齐全，包括柴火壁炉、聚会用燃气壁炉、比萨烤炉、混凝土桌台和户外烧烤炉。

别墅的亮点之一就是由横梁和柱子构成的走廊。它从三车位的车库延伸到寄存室入口，再到独立的客房入口，三者之间的过渡因为这条过道而变得自然完美。

Coachiere 大门既有安保作用，却又十分优雅，就像在欢迎游客的到来。别墅的主卧室不仅具有充分的私密性，宽敞而舒适，还与自然联系紧密，与迷人的沙滩和海洋风景融为一体。总而言之，这是一个开放、友好而亲密的家园。

# Aloe Ridge House

## Aloe Ridge 別墅

**Architects**
Metropole Architects

**Area**
300 m²

**Landscape Design & Implementation**
Dr. Elsa Pooley

**Design Architect**
Nigel Tarboton

**Structural Engineers**
DDR and Associates

**Main Contractor**
IMB projects

**Project Architect**
David Louis

**Design Engineer**
Pat Duffy

**Principal**
Nic Moussouris

**Project Technician**
Simon Wayne

**Project Engineer**
Duran Rammanhor

**Site Foreman**
Ernest Ramekwa

**Location**
Kwa Zulu Natal, South Africa

**Interior Designers**
Dr. Gabriella Lachinger

**Photography**
Grant Pitcher

Under the leafy canopy of an immense Albizia Tree nestles Aloe Ridge House, a 300 m² contemporary home in the Eden Rock Estate on Kwa Zulu Natal's South Coast of South Africa.

The planar estate road (public) facade is intentionally bold, minimalist and austere and hard up against the south western site building line. The result is a visually engaging architecture that makes efficient use of the small site, provides effective privacy to the inhabitants whilst at the same time acting as an efficient barrier to bad weather and prevailing strong winds coming from the south west. In addition a narrow linear plan form, maximizes openness and sheltered private space for living, entertainment and relaxation behind this to the north east, in close proximity to the wild natural bush and looking out towards the view beyond.

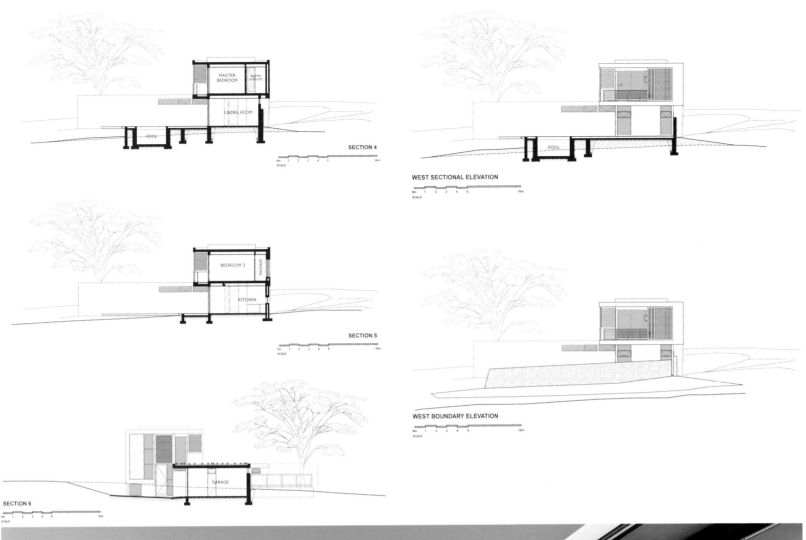

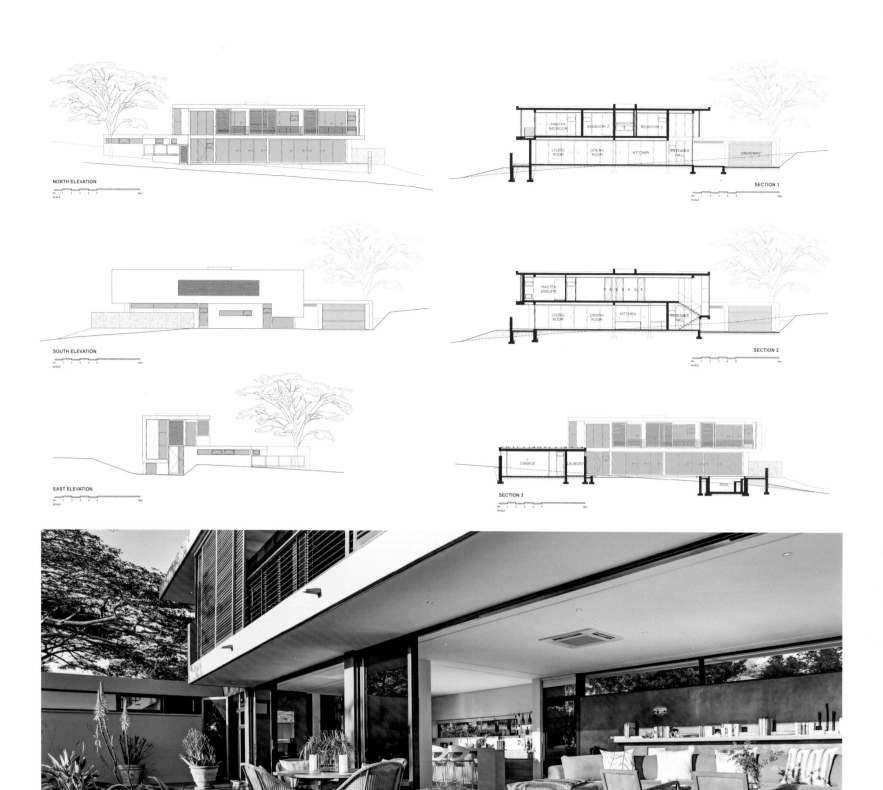

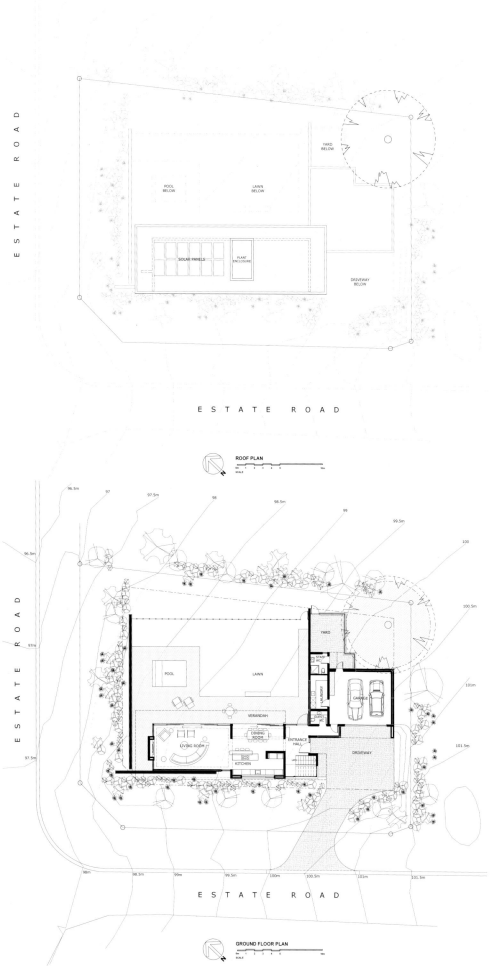

The entrance to the house is a carefully considered grand, double volume arrangement of components in glass, timber and concrete and with "wrap around" form making, a signature characteristic of recent Metropole homes.

There is a sense of "big-ness" and "wow factor" right from the start.

The strong horizontal line created by the roof of the garage structure provides visual axial thrust to the point of entry, into a transparent double volume entrance area and through to the kitchen and living spaces beyond.

Internally, at ground floor level, open plan design with a minimum of dividing walls, no internal doors and level thresholds between inside and outside facilitate a user experience of a single large multi-use space that unconstricted, uncluttered and weather permitting, is able to open up and connect and extend to the outdoors.

High level perimeter strip windows visually lighten the experience of the first floor building mass overhead and enhance the experience of the vertical dimension of the living, dining and entertainment areas at ground floor level.

A generous external decked area with plunge pool and open lawn area beyond encourages the inhabitants to indulge in and celebrate an outdoor lifestyle of entertainment, play and relaxation.

At first floor level, once again the design focus was to promote a sense of openness with privacy and create a diverse, joyful place in a limited space. Whilst the need for privacy has dictated the use of doors, these doorways are full height at 2.6 m and when open allow continuity of space to be experienced through an uninterrupted ceiling plane.

The 3 bedrooms located at this level open out to an elevated balcony which allows distant views over the tree tops to the sea in the east and distant hills and the setting sun to the west. A series of movable Balau timber screens bring in filtered daylight to the clean, modernist interiors, without sacrificing privacy whilst adding a degree of detail and natural colours and texture to the modern facade.

In Aloe Ridge House there is a unity of opposites.

The clean, hard and straight lines of the man-made intervention meet the soft flowing irregular line and textures of the natural bush context in a respectful harmony.

The palette of natural materials including earthy colour tones, timber screens, decking stone cladding juxtapose with the bold and progressive architectural form making, creating a small home that packs a big punch and that is not only visually and spatially exciting, but also comfortable and intimate.

Aloe Ridge 别墅坐落在南非南海岸夸祖卢-纳塔尔省的伊甸石园（Eden Rock Estate）中，四周环绕着茂密的合欢树，占地 300 m²，具有当代建筑风格。

别墅面向园中平坦的公共街道，外观设计大胆，朴素而庄严，沿着西南的建筑线拔地而起。醒目的建筑充分利用了这片小场地，既充分保证了住户的隐私，又充当了防止恶劣天气造成危害的有效屏障，有效阻挡了西南强风。建筑面向东南，狭长的布局中包含了生活、娱乐和休闲空间。这些空间的开放性和隐密性都实现了最大化。别墅十分接近野生灌木，而且可欣赏到壮丽的远景。

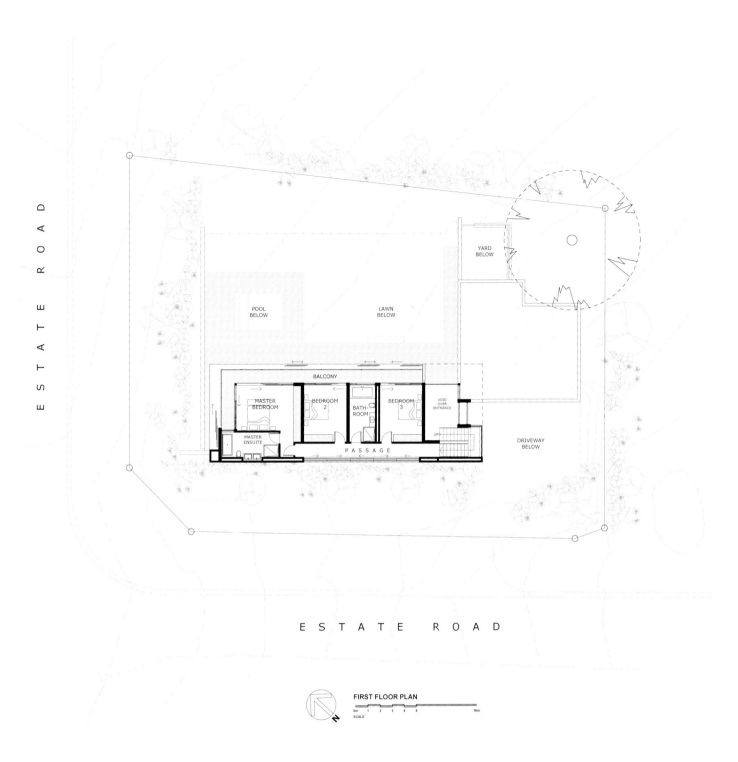

精心设计的房屋入口门厅显得很宏伟,两倍高的空间被玻璃、木材和混凝土覆盖,形成"环绕"式的结构,赋予空间鲜明的都市风格。

从进入房屋的那一刻,你便会因为它的宏伟而感到震惊。

车库的屋顶有着清晰的水平线,与入口处的轴线形成鲜明的对比。再往里走便是两倍高的门厅空间和厨房起居空间。

一楼采用了开放式设计风格,隔离墙很少,也未设内门和门槛,给人一种空间大、功能多的感觉。这个无边际、不凌乱、晴雨皆宜的空间被打开后,便与户外空间融为一体。

宽敞的带状窗户让室内空间瞬间明亮起来,给人以不同的视觉体验。而在客厅、餐厅和休闲空间中,能体验到一种纵向的空间感。

跌水池被宽敞的户外露台围绕着;开阔的草坪也吸引着人们到此玩耍,享受娱乐、嬉戏和放松都在户外的生活方式。

二楼的设计同样注重开放性和隐密性。有限的空间被打造成了气氛愉悦的多元空间。门具有保护主人隐私的功能,全高门廊高达 2.6 m,因此,当门被打开时,室内空间在平坦的天花板下显得颇具连续性。

该层的 3 间卧室都敞向高出地面的阳台。站在阳台上,视线能越过树林,看到东边的大海、远处的群山和西边的日落。一系列的可拆娑罗双木纱窗把日光过滤到装饰清新的现代室内空间中,不仅未影响空间的隐密性,而且为空间的现代外观增添了些许自然色彩。

Aloe Ridge 别墅中既有对立,又有统一。

人为插入的清晰、尖锐、笔直的线条,与柔软、流畅的不规则线条相互交织,在自然树丛的背景中显得十分和谐,令人赞叹。

房屋采用的素色调天然材料包括木纱窗和露台铺石,与大胆、前卫的建筑外形相呼应。小房屋中包含了大空间,不仅在视觉上和空间上令人兴奋,还让人感到舒适、温馨。

# Float House

悬浮小屋

**Architects**
Pitsou Kedem Architects

**Design Team**
Pitsou Kedem, Raz Melamed, Irene Goldberg

**Photography**
Amit Geron

A one storey, private residence in the center of the country. The architectural concept was to create a structure with a continuous, wide space, divided by internal courtyards and movable partitions into smaller spaces used for a variety of different functions.

The different spaces and internal courtyards are joined together into one structure by two, ultra-thin roofs supported at one central point so that seem to float in the air. The two roofs merge, one into the other and extend for 5 m over the building front walls. The entire roof is constructed from lightweight materials and, in order to provide a thin, wispy look at its edges, it is constructed with a moderate slop towards its center.

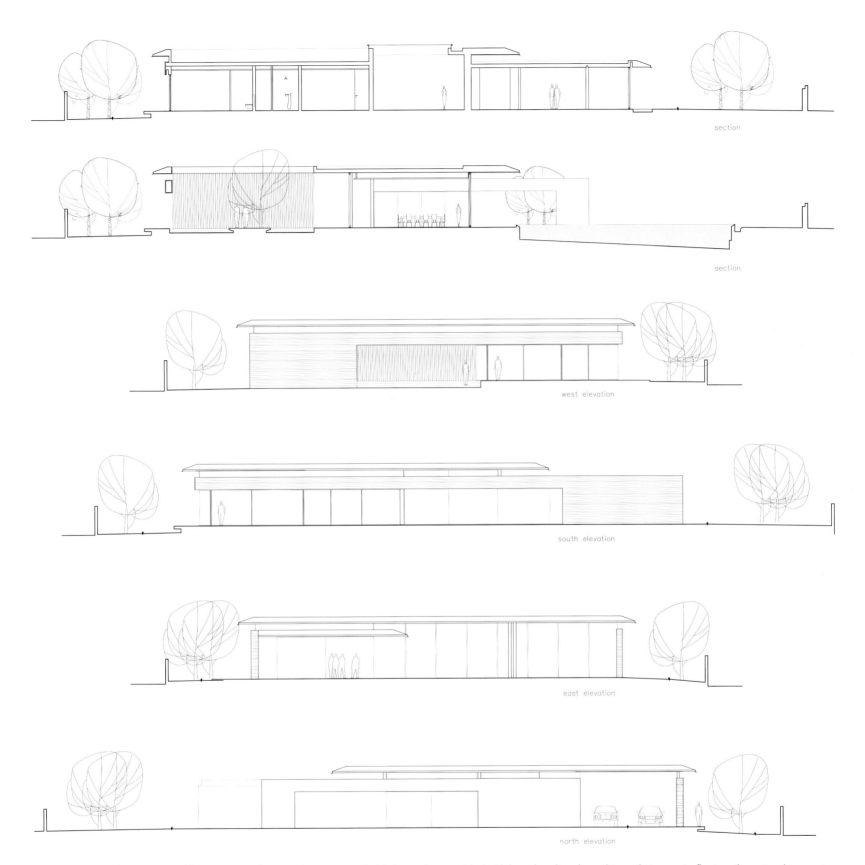

The structure itself is constructed from a series of spaces that are conceivably internal and conceivably external spaces. The entrance is framed with a wall of wooden slats which constitute what could be considered the initial boundary between the outside and the inside.

When entering the space, we pass through a space resembling an entrance lobby — again, conceivably internal and conceivably external — which embodies the soft seam between the outside and inside areas.

Whilst walking through the entrance lobby space, we cross a transparent pool, studded with large basalt rocks and trees that seem to float on the water. As we enter the entrance lobby, we experience the illusion that the house is floating and being reflected, just as the roof appears to be floating above the structures walls.

A ribbon window running along the building's facades serves to emphasize the roof floating above the structure walls and cancels out the feeling of mass that its size suggests.

A long, narrow reflection pool follows the structure's walls, reflecting and emphasizing their covering and texture.

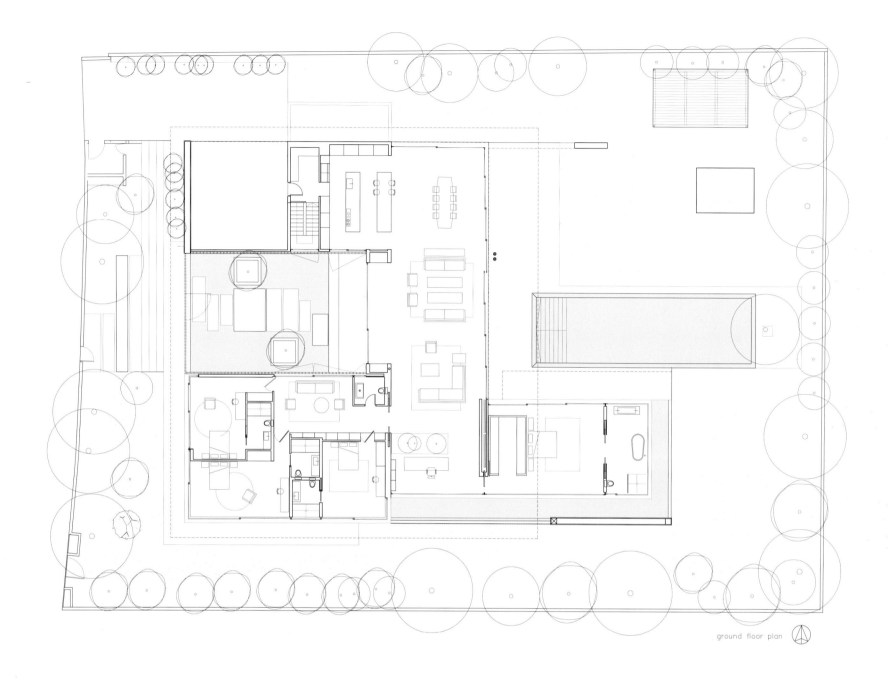

ground floor plan

该单层独栋住宅位于乡村的中心位置。建筑理念旨在打造一个具有连续、宽敞空间的结构，住宅内部使用庭院和可拆除的隔板分隔成各种更小的功能空间。

住宅的各种空间和内部庭院相互连接，形成一个整体结构，两个超薄的屋顶仅用中央的一个支撑点来支撑，从而制造出悬浮在空中的感觉。两个屋顶相互交织在一起，在建筑的前墙处向外延伸了 5 m。整个屋顶都由轻质材料建成，为了形成薄而细的边缘，屋顶向中央稍微倾斜。

建筑结构本身是由一系列的空间组成，包括室内和室外空间。入口处设有由木条组成的门，可以被认为是室内和室外空间的第一条界线。

步入空间，穿过类似于入口门厅的空间。此空间是内外空间的过渡区，令人对内部和外部空间都充满联想。

穿过入口门厅，便能看到清澈见底的水池，树木的倒影和黑色大玄武石仿佛悬浮在水中。步入门厅，人便有种房屋正悬浮于空中的错觉，它与仿佛飘浮在墙壁上方的屋顶有异曲同工之妙。

带状的窗户在建筑的墙壁上延伸，突出了悬浮的屋顶，淡化了因房屋规模产生的堆砌感。

池水从狭长、清澈的水池中流出，沿着建筑的墙壁往下流，凸显了墙壁那有质感的外表。

# Butternut

白托纳特宅邸

**Interior Designer**
Zimenko Yuriy

**Location**
Ukraine, Kiev

**Area**
200 m²

**Photography**
Andrey Avdeenko

The main planning challenge was to combine the two apartments into one. Consequently, the guest area, which occupies about a third part of the total useful floor area, is composed of a combined dining kitchen — drawing room, guest bathroom and a housekeeper room. The rest space is a private area, which is figuratively dimidiated: on the left are rooms of the boys and children's bathroom, across the corridor are: master bedroom, owner's bathroom and walk-in closet. On the right of the entrance room is a small laundry room. Herewith, such a matter, as the plenty of structural columns, the designer has met absolutely in a skillful way: now they are completely invisible. On the site of the balconies sprouted vast bathrooms with panoramic windows: sunniness and extent of overlook regulated by venetian blind.

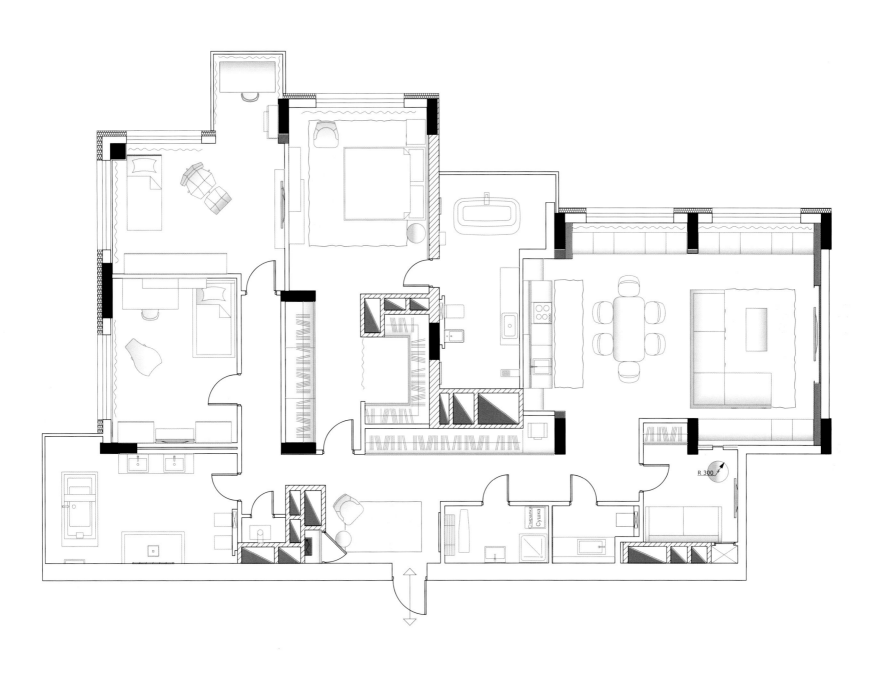

该项目面临的挑战是将两套公寓合成一套。客厅区占用了总面积的三分之一，包括厨房餐间、画室、卫生间和管家卧室。其他空间是被一分为二的私人区：走廊的左侧是男孩们的卧室和浴室，右侧是主卧室、房主浴室和步入式衣柜。门厅右侧设有小洗衣间。设计师巧妙地解决了繁多的结构柱的问题，完全将它们隐藏了起来。朝向露台的宽敞浴室设有全景窗，使空间光线充足，软百叶窗则可以调节视线。

# Oxford 49

## 牛津 49 号别墅

**Project Team**
Adam Court & Tavia Pharaoh

**Interior Decor**
OKHA Interiors

**Location**
Johannesburg, South Africa

**Main Furniture Supplier**
OKHA Interiors

**Photography**
Else young

The JHB home of a French businessman with a love of art & fine entertaining. This 5 bedroom house was designed to be a multi-functional home for family, visiting friends and executive guests. The brief was to express an individual ambient character but maintain a calm and uncluttered elegance. Every vase, sculpture, artwork and furniture object is given enough meditative space to allow it to be seen and appreciated. Interiors by Antoni Associates.

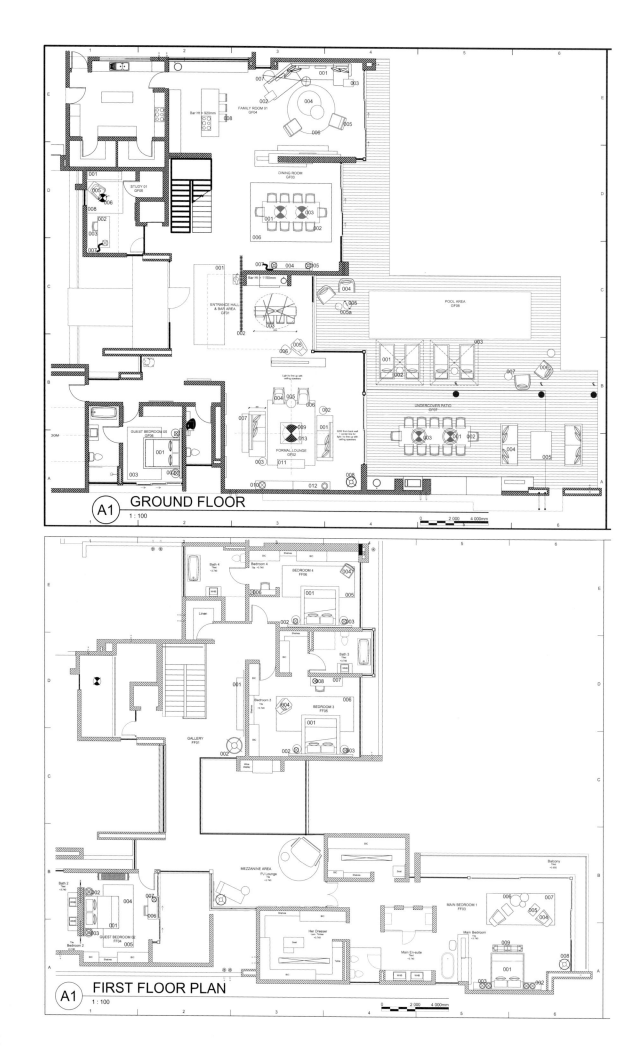

/ 213

该别墅位于约翰内斯堡,为一位法国商人所有。业主热爱艺术和休闲娱乐。这栋多功能家庭住宅的设计考虑到家人、朋友和客人的到访,设有 5 间卧室。设计旨在展示独特的环境特征,但同时营造平静、整洁和优雅的氛围。每件花瓶、雕塑、艺术品和家具都在视线内,供人欣赏,给人带来无限的遐思空间。室内设计由 Antoni Associates 公司设计。

# Wallace Ridge

华莱士山脊别墅

**Architects**
Whipple Russell Architects

**Location**
Beverly Hills, California

**Area**
622 m²

This project began with former clients' wish to move back closer to the city of Beverly Hills. They had found a potential property in Trousdale Estates and showed it to Marc for his advice on its possibilities. The property was in disrepair, had a choppy floor plan, and gabled roofs that did not fit the client's vision of a modern home. The goal, of course, was to maximize the views while creating fluid well-lit spaces that would both serve and reflect the lives of the inhabitants. Marc saw a way to stay within the Trousdale Estates' single storey 4.3 m height restriction and still provides spaciousness and a spectacular view.

The entire core of the house was redesigned to feature an open plan, high ceilings and a sleek flat roof. The front door, flanked by large glass panels opens to a wide entry a perfect stage for a piano and provides an open sight line across the living area, though 3.7 m high glass walls and 2.4 m high glass sliding pocket doors, to the patio and pool.

The clients do a lot of entertaining and required a kitchen that was open to the living and outdoor spaces. Rooms are minimally defined using tall panels, custom stained in a rich coffee bean brown, that contrast with the light walls, spaces suited to artwork. The warm modernism the clients wanted was achieved with a harmonious use of materials as kitchen, dining, entertaining and living room spaces flow easily into one another. In the living area a large screen television and fireplace are recessed into wall-sized expanse of Portico Slate tile by SOLI. In the kitchen, the island and countertops are Caesar stone in Lagos Blue and cabinets have an acrylic lacquered finish.

Large glass pocket doors open to the outside from both kitchen and living areas, where there is a patio bar, conversation areas and a tabletop fireplace all encircling the pool. The master suite also opens to the pool though large sliding glass panels. To create a vantage point for the best view, a roof terrace was built atop the master suite, accessible from the pool area. The master terrace provides space for entertaining, sunbathing, a game of table tennis, and a view all the way to the Pacific. Below, the master wing offers a library/sitting room, and a home theater.

Next to the bedroom, the master bath continues the use of brown with the tile in the master shower — a basket weave pattern from SOLI. Adjacent is a roomy closet/dressing area. As there is a musician in the family, the clients wanted to find space for a full music studio; Marc found it by digging down and locating the studio beneath the motor court. It includes a separate control room, sound booth for vocal recording, and tracking room, a soundproofed oasis for creation.

按照客户想就近贝弗利山而住的愿望，该项目开始拉开帷幕。客户在特劳斯代尔庄园找到了一处很有开发潜力的房产，并向马克询问了有关房子建设可能性方面的建议。这栋房子多年失修，楼板起伏不平，尖尖的人字形屋顶更是不符合客户对现代房子的要求。设计目标是打造流畅明亮的空间，能够表现居住者的生活方式，同时，使屋内视野最大化。特劳斯代尔庄园对单层房屋的高度限制是 4.3 m，马克想出了一个方法，既能将房屋高度控制在这一范围之内，又能使空间宽敞，且住户能欣赏到户外壮丽的风景。

设计师重新将房屋内部设计成开放式空间，其中包括高高的天花板和平坦的屋顶。宽敞的正门两侧都是玻璃墙，使得屋内布置一目了然，一架钢琴正好摆放在这个完美舞台中央。透过 3.7 m 高的玻璃幕墙和 2.4 m 高的滑动玻璃折叠门，就可以看到屋外的露台和水池。

客户日常招待活动较多，因此要求厨房与客厅及户外空间相连。高高的墙板最低限度地划分了空间。墙板被漆成咖啡色，与适合挂艺术品的白色墙壁形成对比。柔和的现代风格的实现得益于材料使用的协调一致，如厨房、餐厅、娱乐和客厅空间都采用了相同的材料，使得空间之间过渡自然，十分流畅。客厅中的大屏电视和壁炉设在一起，都嵌入到镶有 SOLI 门廊石板砖的墙壁之中。厨房设有独立灶台，台面是拉各斯蓝恺萨金石，而橱柜外表都涂有丙烯酸漆，十分精致。

推开厨房和客厅的大玻璃折叠门，可直接通往围绕水池而设的户外活动空间，包括户外灶台、露天吧台和交谈空间。主卧室的大滑动玻璃门也敞向水池。为了获取良好的观景视角，主卧室的楼顶设立了露台，从水池区就可直接登上露台。露台上拥有娱乐、日光浴、乒乓球设施，还能欣赏太平洋远景。下方是房屋的一翼，设有书房（休息室）和家庭影院间。

卧室旁边的主浴室同样采用了棕色色调，主淋浴区铺满了棕色的编篮纹 SOLI 瓷砖，隔壁是大橱柜（更衣空间）。由于家庭成员中有一位音乐家，因此客户想打造一间音乐工作室。马克把该空间设定在停车场下方，通过往地下挖，为独立的控制室、录音隔音室和录音棚腾出了足够的空间，另外还设有为创作打造的隔音空间。

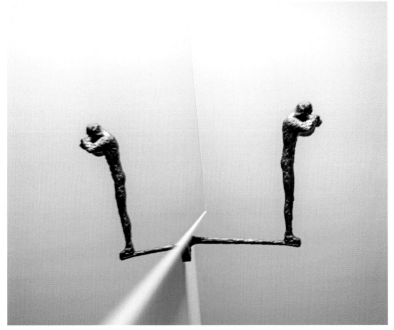

# Clovelly House

克劳夫利别墅

**Architects**
Rolf Ockert Design

**Location**
Sydney, Australia

**Client**
Private

**Builder**
Tony Kerle - SL Wilson

**Photography**
Sharrin Rees

The design process, complex due to several defining key issues inherent to the site, ended up taking us through some radically different sketch options before settling on the one that was finally pursued.

These key factors together with of course countless smaller factors and decisions along the way shaped the house to what it is. The unusual but elegant roof shape allows sunlight in while still allowing neighbours to enjoy water views over the lower end. The expressive angled concrete wall mirrors the roof shape but in negative, resulting in complex facade geometry along the main face, enhanced by the movement of ever changing shadows over the shapes.

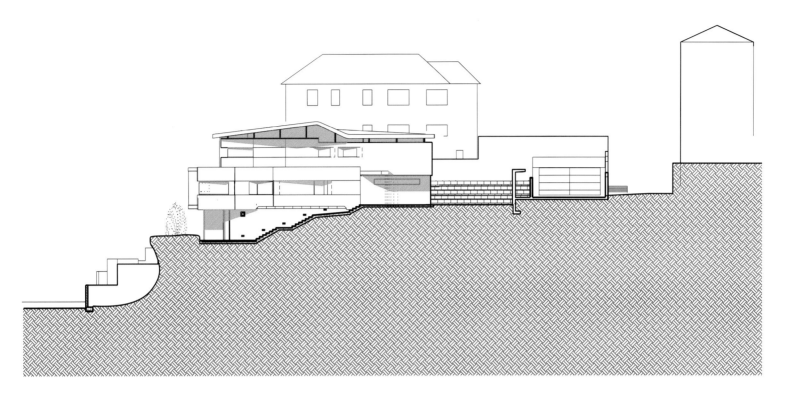

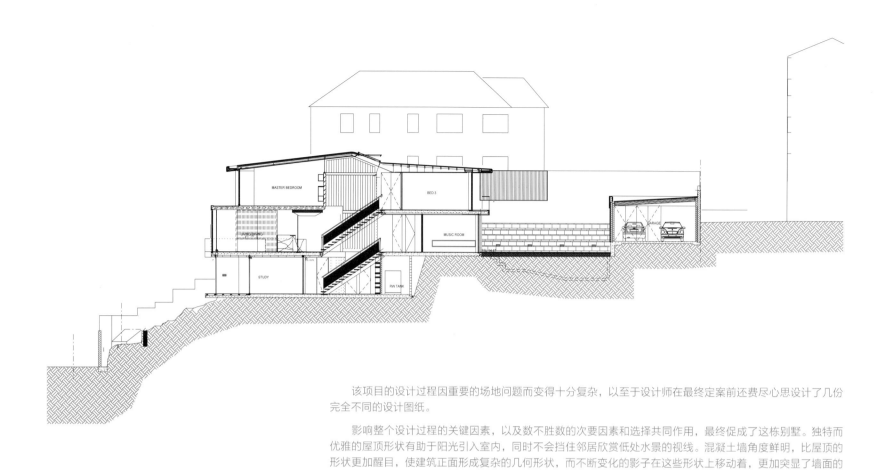

该项目的设计过程因重要的场地问题而变得十分复杂,以至于设计师在最终定案前还费尽心思设计了几份完全不同的设计图纸。

影响整个设计过程的关键因素,以及数不胜数的次要因素和选择共同作用,最终促成了这栋别墅。独特而优雅的屋顶形状有助于阳光引入室内,同时不会挡住邻居欣赏低处水景的视线。混凝土墙角度鲜明,比屋顶的形状更加醒目,使建筑正面形成复杂的几何形状,而不断变化的影子在这些形状上移动着,更加突显了墙面的几何效果。

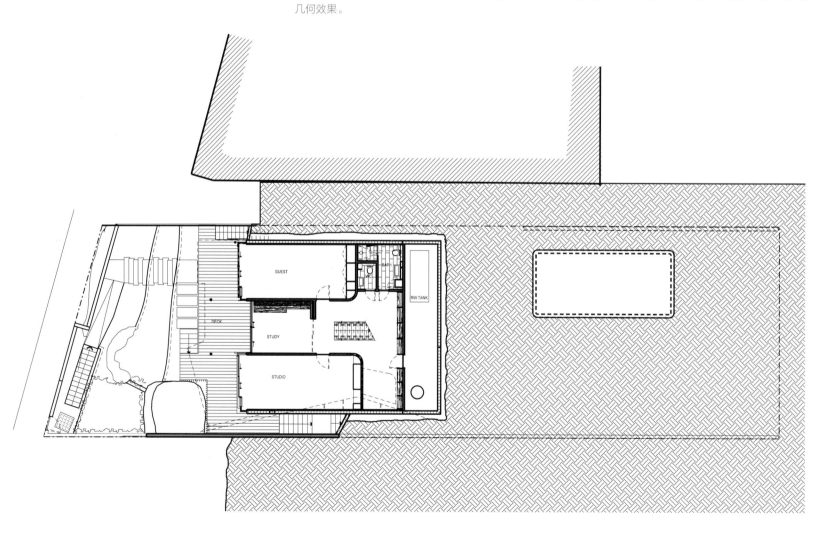

The light void also contains the central circulation, the stairs. These are light and airy without looking or feeling flimsy. To the north of are 2 levels, to the south 3, taking advantage of the natural slope of the site. The main living space is on the entry level, connecting it with the northern garden and pool as a very generous central family area. Upstairs are the bedrooms, on the southern lower level several areas for more individual activities, study, studios and library.

As a consequence of the relentless southerly winds the house was designed, unusually and against our original instinct, without any opening windows facing south. Instead large frameless floor-to-ceiling double glazed elements allow uninterrupted views over the Pacific and allow a more intimate visual connection than framed openable glazing elements would have afforded.

An outdoor deck is attached to the side of the living area, allowing outdoor activity on suitable days without interruption of the front row feel the house enjoys.

The original southern slope in front of the house was full of building rubble from some previous building incarnation. Once that was all removed several large natural sandstone blocks that had fallen eons ago and stood upright, affording us an unexpected, giant Japanese-style rock garden.

The materials for the house were chosen for a variety of reasons. First of all they had to be suitable for the harsh salt spray environment where everything gets a thick coating of corrosive salt within a few days. They were also desired to be suitable for the location, reminiscent of flotsam, rich but weathered. Finally, the rich natural palette of coastal colours, grey and red in the rocks, blues and greens in ocean and sky, provided already a magnificent canvas of hues and textures dominating large parts of the house. The rich geometry in the house as well as the resulting ever changing play of light in the interior and exterior spaces also meant that we did not feel the need for strong colour or texture, elements we often love to employ in other settings.

The resulting material palette relies on very few elements, the strong raw concrete, along the outside wall as well as to the living room ceiling. A dark Zinc roof, being allowed to weather. Dark timber in floors and joinery, both offset with white walls and joinery faces. And of course the ubiquitous glass. Both, outside floors and walls as well as interior benches are finished in the same, earthy grey stone. Having these few elements used in a diverse range of applications throughout the house also helped to tie the many different spaces of the house together to a coherent whole.

Environmental concerns also played a big role in the development of the design. The aforementioned central void allows natural light deep into the heart of the house, eliminating the need for artificial lighting during daytime.

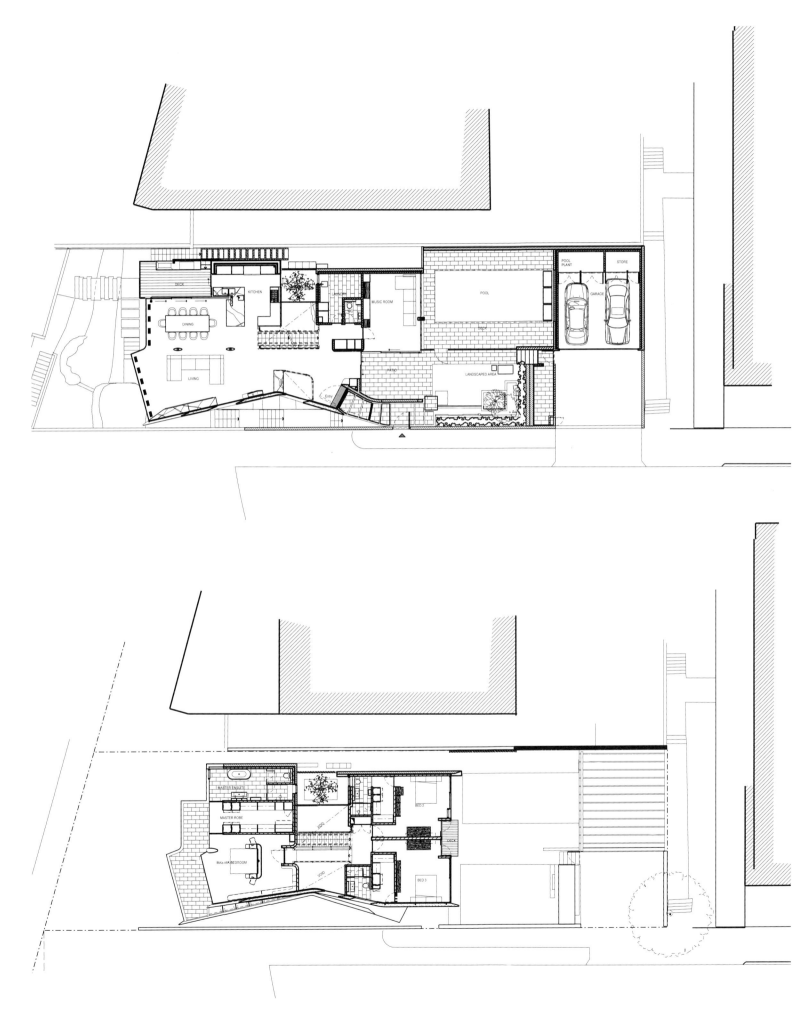

This void, supported by the roof shape in combination with the operable skylights, also helps to naturally ventilate the entire house as it allows the rising hot air to escape, drawing cooler air behind. Operable floor vents in the living area allow for the ubiquitous sea breeze to be let into the house in a controlled manner, all but eliminating the need for air conditioning.

High performance insulation and double glazing throughout in combination with the high thermal mass in the house allow for utilisation and storage of the northern solar heat gain in winter, keeping the house warm during the colder months.

Large rainwater storage tanks are sufficient to fill the pool and water the indigenous planting throughout the grounds.

All lights are low wattage LED type, reducing the electricity use significantly.

空间内还有着中央流通空间和楼梯，明亮通风，也十分坚实。建筑朝北的方向设有2层，而朝南的方向设有3层，充分发挥了场地的坡度优势。主起居空间设在1层入口处，与北边的花园和水池相接，空间十分宽敞，充当了中央家庭空间。

为了保护建筑免受严酷的南风侵蚀，房屋南墙未设任何窗户，颠覆了人们往常的认识。而其他部分则采用两倍高无框玻璃幕墙，代替了带框可开玻璃窗，使得室内外在视觉上更加连通，从室内远眺太平洋的风景也丝毫不受阻碍。

户外平台的一侧与客厅区相连，当天气适宜时，就可以进行户外活动，同时也并不影响房屋的有序感。

房屋南面的斜坡上原来堆满了以前的建筑碎石，经过搬移，几块远古时期坠落的巨大自然砂石屹立在了园中，形成了一个日式的大型假山庭园，令人惊叹不已。

材料选择遵循了多样化的依据：首先，材料必须适应严酷的烟雾环境，因为在这样的环境中，所有的东西在短短的几日就会附上一层厚厚的腐蚀盐；其次，材料必须符合场地的风格，让人联想到丰富却又饱经风霜的漂浮物；最后，材料必须具有丰富、自然的海岸色彩，比如灰色和红色的岩石、海洋和天空的蓝色和深绿色，让大部分空间都充满丰富的色彩和非凡的质感。房屋那丰富的几何形状，以及室内和室外空间光线强弱的不断变化，让我们感觉空间中无需再加入任何色彩、质感，然而，这些正是设计师通常喜欢应用到陈设中的元素。

因此，材料选择很少，建筑外墙和天花板都是坚固的混凝土。深色的镀锌屋顶经得住天气的侵蚀。地面和木工品都采用了深色木材，对白色的墙壁和木制品的外观起到了互补的作用。此外还有随处可见的玻璃材料。户外的地面和外墙与室内的长椅有着相同的大地灰色外观。虽然材料很少，但却应用到了房屋的各个角落，用途广泛。同时还将房屋的不同空间紧紧地连在一起，使空间更有整体感。

对环境的考虑在设计及其执行过程中也扮演着重要的角色。上述的中庭能把自然光线引入房屋的深处，因此，白天时房屋无需采用灯光来照明。

中庭因屋顶结构而更有空间感，配上可开关的天窗，空间更加明亮通风，热空气上升到此处时可以直接排到屋外，而更多凉风将进入室内。客厅区地面的可开关通风口可控制引入室内的海风，因此，室内无需用空调来调节室温。

被应用到房屋各处的高性能双层绝缘玻璃以及热量收集设施，有利于房屋收集并利用冬季阳光和从北方照射过来产生的太阳热，使得房屋在寒冷的月份时也有较强的保暖性。

大型蓄水池能够收集足够的雨水，来补给水池用水和整个场地的植物用水。

房屋采用的全是低功率LED灯，大大减少了用电量。

# Kona Residence

科纳别墅

**Architects**
Belzberg Architects

**Interiors**
MLK Studio

**Principal**
Hagy Belzberg

**Area**
725 m²

**Project Manager**
Barry Gartin

**Location**
Kona, Hawaii

**Project Team**
David Cheung, Barry Gartin, Cory Taylor, Andrew Atwood, Chris Arntzen, Brock DeSmit, Dan Rentsch, Lauren Zuzack, Justin Brechtel, Phillip Lee, Aaron Leppanen

**Photography**
Benny Chan (Fotoworks), Belzberg Architects

Nestled between cooled lava flows, the Kona residence situates its axis not with the linearity of the property, but rather with the axiality of predominant views available to the site. Within the dichotomy of natural elements and a geometric hardscape, the residence integrates both the surrounding views of volcanic mountain ranges to the east and ocean horizons westward.

The program is arranged as a series of pods distributed throughout the property, each having its own unique features and view opportunities. The pods are programmatically assigned as 2 sleeping pods with common areas, media room, master suite and main living space. An exterior gallery corridor becomes the organizational and focal feature for the entire house, connecting the two pods along a central axis.

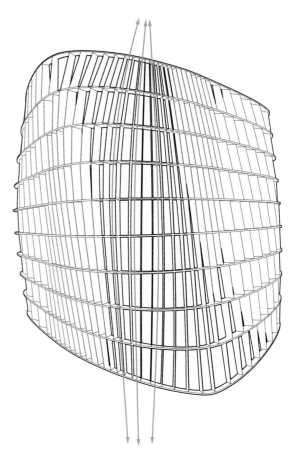

To help maintain the environmental sensitivity of the house, 2 separate arrays of roof mounted photovoltaic panels offset the residence energy usage while the choice of darker lava stone help heat the pool water via solar radiation. Rain water collection and redirection to 3 drywells that replenish the aquifer are implemented throughout the property. Reclaimed teak timber from old barns and train tracks are recycled for the exterior of the home. Coupled with stacked and cut lava rock, the 2 materials form a historically driven medium embedded in Hawaiian tradition. Local basket weaving culture was the inspiration for the entry pavilion which reenacts the traditional gift upon arrival ceremony. Various digitally sculpted wood ceilings and screens, throughout the house, continue the abstract approach to traditional Hawaiian wood carving further infusing traditional elements into the contemporary arrangement.

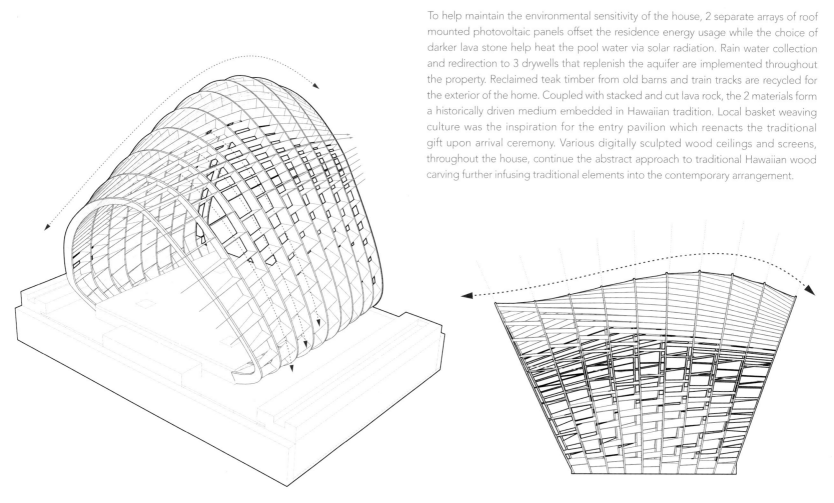

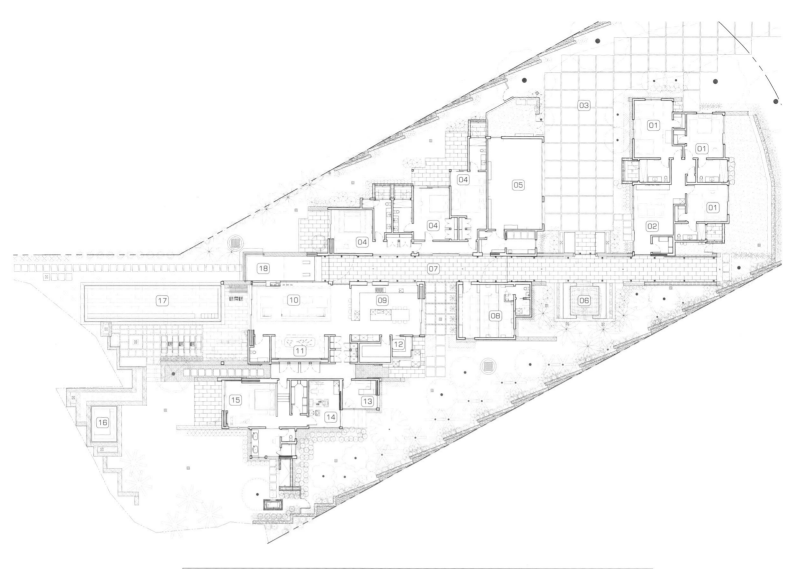

| | | | |
|---|---|---|---|
| 01 KID'S ROOM | 06 ENTRY PAVILION | 11 DINING ROOM | 16 HOT TUB |
| 02 KID'S COMMON | 07 OUTDOOR GALLERY | 12 OFFICE | 17 POOL |
| 03 MOTOR COURT | 08 THEATER | 13 OFFICE | 18 REFLECTING POOL |
| 04 GUEST ROOM | 09 KITCHEN | 14 GYM | |
| 05 GARAGE | 10 GREAT ROOM | 15 MASTER BEDROOM | |

科纳别墅坐落在冷却的熔岩流之间,纵向的布局巧妙地获取了横向延伸所没有的非凡风景。住宅融合了自然元素和几何硬景观两大特征,引入了东边的火山风景以及西边的海洋景色。

排列整齐的结构分布在建筑的各个角落,就像一系列的"豆荚"。然而,每处却各有特色,各有风景。这些豆荚被划分成各种区域,如2间带有公共空间的卧室、媒体室、主套房和主起居区。室外的走廊贯穿了整栋房子,具有组织作用,连接了主轴线两侧的区域,成为了房子的特色。

为了维持房屋对周围环境的敏感度,屋顶上装有2排太阳能板,给房屋供应能源。黑色的熔岩石在太阳辐射下散发出热量,给池水加热。别墅通过3个旱井来收集和分配水,补给整栋建筑的蓄水层。再生柚木取自旧畜棚和火车轨道,被重复利用到户外空间中,再配上经过堆叠、加工的火山石,使得别墅颇具夏威夷传统特色。入口处亭子的灵感则来自于当地的篮子编织文化,就像欢迎仪式上的"传统礼物"。各式各样的数码雕刻木天花板和屏风渗透到了整栋房屋,延续了传统夏威夷抽象木刻手法,此外,传统元素也与当代的布置风格相融合。

# House Boz

博兹别墅

**Architects**
Nico van der Meulen Architects

**Designer**
Werner van der Meulen

**Location**
Mooikloof Heights, Pretoria, South Africa

**Residence Size**
777 m²

The client requested a spacious and luxurious four bedroom house with an emphasis placed on the design of the living rooms. Ensuring that the magnificent views were optimized was of utmost importance, the design of this 777 m² house not only responds well to the client's requirements but also to the context of the site.

Every design decision communicates and reinforces the concept, as can be seen in the selection of materials used and the way the internal spaces relate to the outdoors. With the choice of materials predominately natural materials and earthy colors, it is evident that even the smallest of details make reference to the concept in a very unique way. Initially the site revealed itself as a mound of quartzite rock which was excavated and hand cut for the gabion walls and the stone cladding used throughout the house.

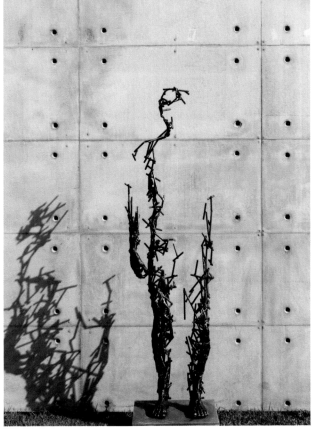

/ 245

The entrance hall positioned between the double garages is distinctively located alongside a partially covered atrium that gently introduces you to several views through the house as you're welcomed in. The koi pond introduces water as you approach the front door while various podiums add depth and dimension to this space. An elevated sculpture podium and interlocking planters bring this atrium to life. This also makes it possible to sleep with the doors open as it is impossible to get into the atrium once the Mentis grating gate to the driveway is locked.

The double volume entrance hall is framed by a back-lit perforated skin of scaffolding boards bolted to a wall, creating the perfect backdrop for the sculptural looking concrete staircase with steel inlays. Underneath the staircase is a sculpture by Regardt van der Meulen.

The kitchen overlooks the lanai and garden while the frameless folding doors create an invisible threshold between the inside and out. These doors, when completely open, allow for the kitchen and dining room to overflow onto the lanai and bar, making entertaining effortless and enabling adults to keep an eye on children in the pool, a mere meter away.

The lanai with a sunken jacuzzi is snugly positioned between the pool on one side and a stone-cladded wall on the west which screens the afternoon sun to ensure the lanai's temperature remains moderate. It is these design decisions that truly set this house apart from the rest.

All four en-suite bedrooms are situated on the first floor with all of the bedrooms having their own private balcony. The 3 children's bedrooms are situated on the western wing of the house while the main bedroom is located on the eastern wing. A suspended walkway with steel sheeting as floor tiles overlooking the atrium links the 2 wings and creates a sense of privacy for the main bedroom.

Phia van der Meulen and the M Square Lifestyle Design team strategically linked spaces through their use of various natural materials in the interior spaces. In situ-concrete, quartzite cladding and rusted mild steel were incorporated into the design.

The interiors feature linear and monolithic forms that complement the architect's vision for this house. Many of the functional elements were designed to become beautiful features that visually connect the spaces and create links throughout the house rather than just remaining purely functional. An example of this would be the way the staircase relates to the aluminum ceiling which features in both the main living room as well as in the main bedroom.

Regardt van der Meulen's original steel sculptures were chosen for the project, as they fitted perfectly with the steel theme of the project. The selection of furniture pieces once again continued this theme where splashes of orange were used in the living room making reference to the orange seen in the rusted metal cladding. The overall charcoal colour range used in this house complements the shades of grey of the concrete walls.

设计公司应客户的要求,设计一栋宽敞而豪华的四居室别墅。此外,客户还特意强调了客厅的设计。该别墅占地面积 777 $m^2$,设计的重点是保证别墅内视野最佳化,同时满足客户的要求,实现房屋与环境的融合。

项目的每个设计决定都传递和强调了设计理念。这不仅体现在材料选择和使用上,还体现在室内空间与室外空间的连接方式上。在最细微的细节之处也都能体现设计理念。项目主要采用朴素的大地色天然材料。场地本来堆满了石英石,经过挖掘和人工切割,被应用到石笼墙上。房屋各处都以石头装饰。

门厅设在两个车库之间,位于部分被遮盖的中庭旁,位置十分突出。从门厅进入便可看到房间内的几处景观。穿过前门就可观赏锦鲤鱼池。装饰着空间的各种矮墙,使空间更有深度和规模。锁上朝向车道的栅门,就无法进入中庭,因此,敞开中庭的大门睡觉都很安全。

两倍高门厅构架的形成得益于脚手架板的使用。脚手架板被螺栓固定在墙壁上,背光灯的灯光可以从板面的穿孔照射出来,恰好成为了用钢筋水泥组成的大楼梯的背景。楼梯下面摆放着南非艺术家 Regardt van der Meulen 的雕塑作品。

从厨房可以俯瞰门廊和花园,屋内的无框折叠门是一道无形的门槛,划分了室内与室外空间。当折叠门被完全敞开,厨房、餐厅与阳台、酒吧融为一体,不费吹灰之力,便营造出娱乐空间,此处离泳池很近,仅有一米之隔,大人可以在屋内看到在泳池中的孩子。

阳台隐藏在水池和西石墙之间,并设有按摩浴缸。墙壁为阳台遮住了午后的阳光,使阳台温度适中。正是这种设计让房子与众不同。

4 间卧室套房都设在一楼,每间房都带有独立阳台。3 间儿童房位于房子的西翼,主卧室则位于东翼。悬浮的走道采用钢板做地板砖。在过道上可以俯瞰连接两翼的中庭,也给主卧室增添了隐密性。

Phia van der Meulen 建筑设计公司和 M Square Lifestyle Design 设计团队用各种天然材料巧妙地把室内空间连为一体。另外,现浇混凝土、石英石覆盖层和生锈的软钢也都被应用到设计中。

完整线型的室内形式,使房屋比建筑师原来期望的更加完善。许多功能元素通过设计,变成了美观的室内陈设。它们从视觉上将空间连为一体,并建立起空间之间的联系,而非纯粹停留在功能性上。楼梯与铝板天花板的连接方式很好地体现了这一点。而且,主客厅和主卧室内都采用了铝板天花板。

Regardt van der Meulen 的钢塑像,恰好符合了项目的主题。家具的选用也遵循了这一主题,客厅内喷溅了橙色,与生锈的金属层颜色相似。房屋整体采用了炭黑色调,与灰色的水泥墙互补。

# Residence Amsterdam

## 阿姆斯特丹住宅

**Architects**
Robert Kolenik

**Area**
450 m²

This project, a 450 m² villa, is located in the north of the Netherlands. With the owners throwing a house warming party to welcome everybody in their new home, the villa is now definitely ready. Robert placed a few outstanding eye catchers in the interior. The Maretti "Dream" chandelier designed by Kolenik for instance. Situated in the basement are a wellness room and a bar. De master bedroom has a view on the "Dream" Chandelier. Simply pushing a button can easily blind that view. Innovative privacy glass makes this possible.

　　住宅坐落于荷兰北部，占地 450 m²。房屋落成时，业主在房屋内举办了新居落成派对。罗伯特的室内设计有几处亮点，如 Maretti "梦幻"吊灯；地下室中则设有健身房和酒吧。主卧室处能看到"梦幻"吊灯，但当按下按钮时，玻璃落下便营造出私密的空间，这得益于创意墨色遮光玻璃的使用。

# Seacliff Residence

锡克里夫别墅

**Designer**
Paul McClean

**Location**
Los Angeles, CA

**Photography**
Nick Springett

A long and difficult planning process added to the challenge of designing this home. A restrictive height limit plus views from neighboring properties that needed to be preserved on all sides limited the location and size of the upper floor. This resulted in a 3-level home with primary living spaces on the top floor, entry and master bedroom at the ground floor and additional bedrooms below with views out over a steep slope in the rear yard. Because space was limited, the main living area was streamlined into one useable area with the kitchen separating dining from living. The kitchen itself consists of a full height cabinetry wall in white glass with a glacier white marble island with ample seating and cooking space. The room is wrapped in glass on 3 sides to increase the feeling of volume and sliding Fleetwood doors fold back to reveal a large entertainment deck with lounge and BBQ areas.

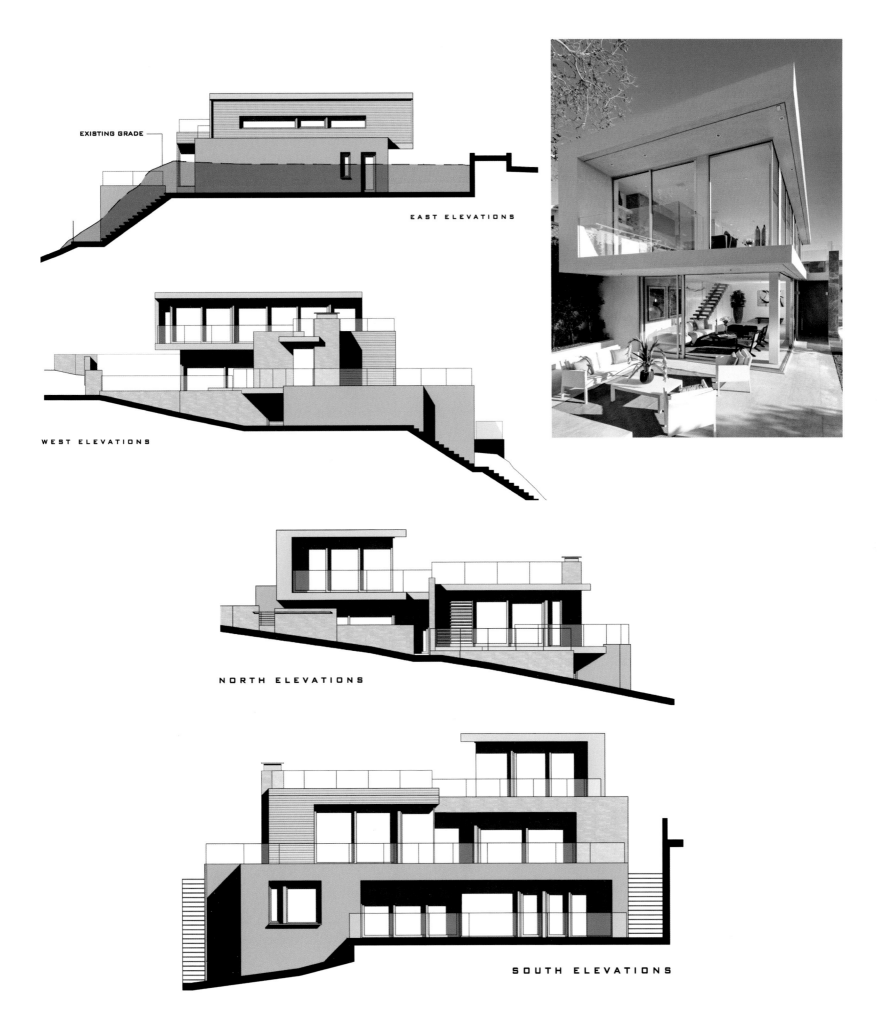

EXISTING GRADE

EAST ELEVATIONS

WEST ELEVATIONS

NORTH ELEVATIONS

SOUTH ELEVATIONS

/ 259

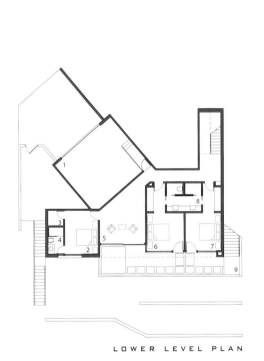

LOWER LEVEL PLAN

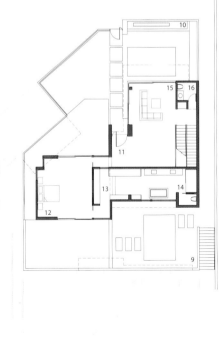

MAIN LEVEL PLAN

UPPER LEVEL PLAN

## LEGEND

| | | |
|---|---|---|
| GARAGE 1 | PATIO 9 | DECK 17 |
| BEDROOM 1 2 | WATER FEATURE 10 | DINING ROOM 18 |
| CLOSET 1 3 | ENTRY 11 | KITCHEN 19 |
| BATHROOM 1 4 | MASTER BEDROOM 12 | FAMILY ROOM 20 |
| LOUNGE 5 | M. CLOSET 13 | |
| BEDROOM 2 6 | M. BATHROOM 14 | |
| BEDROOM 3 7 | LIVING ROOM 15 | |
| BATHROOM 8 | POWDER 16 | |

SCALE IN METERS

NORTH

The entry is approached through a walled off courtyard and leads directly into a family room focused on the courtyard and its water feature. The master bedroom has views to the waves below and a unique combined closet and bath with beautiful Italian closets and fixtures. At the lower level the 3 bedrooms and family area open onto a terrace with views of the garden below. The palette of the house is designed to reflect its beach side location, light walls and soft limestone and oak accented with glass.

漫长而艰难的策划过程增加了项目的难度。为了不影响周围建筑的视野，此建筑四周都有一定的空间，使得建筑高度、顶层位置和大小都受到了限制。因此，房屋只建了 3 层，顶层仅设有主要的起居空间，而大厅和主卧都设在一楼，其他的卧室朝向后院，能够欣赏到斜坡上的风景。由于空间有限，主起居区呈线型延伸，各功能区融为一体，厨房把餐厅从起居区分隔开来。厨房中设有全高白色玻璃橱柜墙，白色大理石厨房灶台提供了充足的席位和烹饪空间。空间的 3 面都被玻璃覆盖，增加了房屋的空间感；推开滑动弗利特伍德木门，便来到了宽敞的娱乐区，休息区和野外烧烤区都设在此处。

步入环绕着围墙的庭院，就是家庭公共空间，被水景装饰着。从主卧室往外眺望，就能观赏微波粼粼的水池，空间还设有别致的组合式橱柜，浴室中则装有意大利橱柜和灯具。楼下是 3 间卧室和公共区域，空间敞向露台，不远处便是花园。房屋的用色方案旨在体现滨海位置特点，浅色的墙壁、暖色调石灰岩以及项目都与空间的玻璃相呼应。

# Albizia House

合欢屋

---

**Architects**  
Metropole Architects

**Design Architect**  
Nigel Tarboton

**Project Architect**  
Tyrone Reardon

**Project Technician**  
Chris Laird

**Structural Engineers**  
Young & Satharia

**Design Engineer**  
Rob Young

**Structural Technician**  
Terry Schubach

**Interior Designers**  
Union 3 Clifton Smithers

**Main Contractor**  
East Coast Construction

**Principal**  
Justin Rosewarne

**Project Manager**  
Benno Terblanche

**Site Foreman**  
Tony Moodley

**Site Area**  
4,360 m²

**Building Area**  
1,000 m²

**Photographer**  
Grant Pitcher

---

We were commissioned to design a contemporary family home on a one acre site, situated at the end of a spur, in Simbithi Eco-Estate. The clients brief called for a home with an overriding sense of simplicity but with a high degree of sophistication.

All the living areas and bedroom suites face onto a panoramic vista, which includes a dense forest down-slope from the house.

The palette of natural materials including timber screens, decking and cladding, off-shutter concrete and stone cladding juxtapose with the aggressive architectural form making, creating a home that is not only visually and spatially exciting, but also comfortable and intimate.

The extensive use of water in the design of the home includes a 25 m lap pool with a

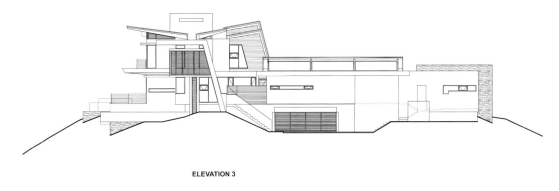

**ELEVATION 3**

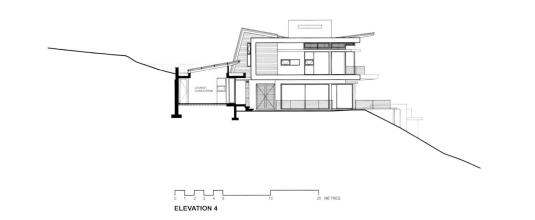

**ELEVATION 4**

glass panel between the water and the basement cinema room, and a shallow but expansive reflective pond on the approach side, which mirrors the building day and night, and evokes a sense of tranquility.

The architectural style of the home is heavily influenced by the "Googie" architecture of the American architect John Lautner. The origin of the name "Googie" dates to 1949, when architect John Lautner designed the West Hollywood coffee shop, Googies, which had distinct architectural characteristics.

"Googie" architecture is a form of modern architecture and a subdivision of futurist architecture with stylistic conventions influenced by, and representing 50's American society's fascination and marketing emphasis on futuristic design, car culture, jets, the Space Age, and the Atomic Age.

"Googie" was also characterized by design forms symbolic of motion, including upswept roofs, curvaceous geometric shapes, and the bold use of glass, steel and neon, the spirit of which is embodied in Albizia House.

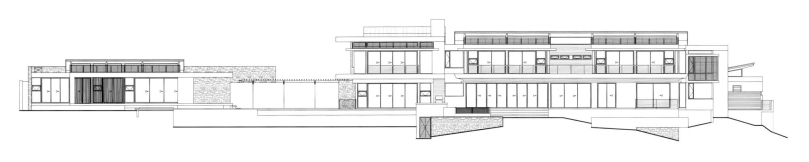

**ELEVATION 1**

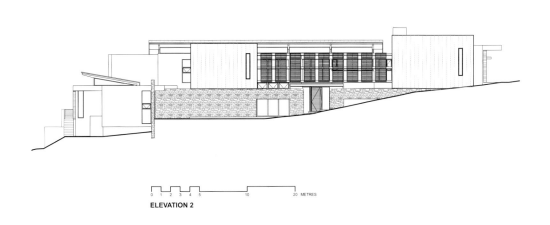

**ELEVATION 2**

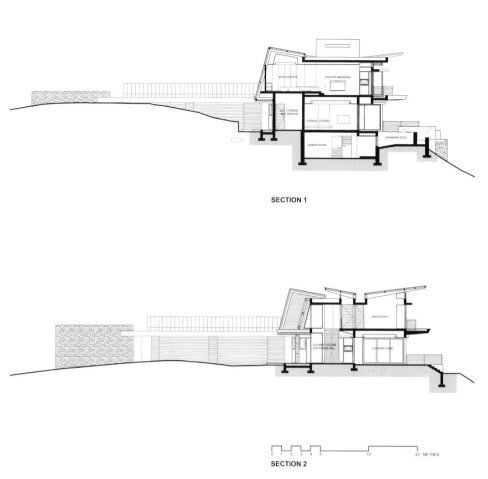

SECTION 1

SECTION 2

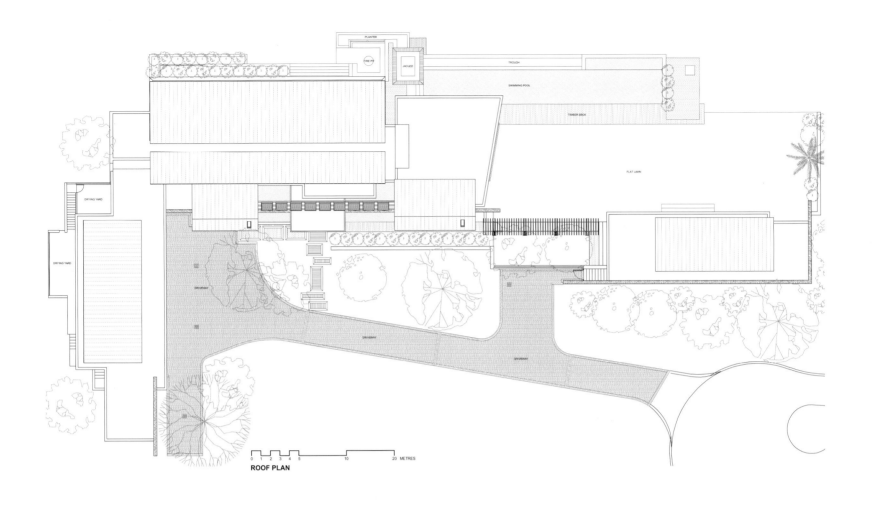

ROOF PLAN

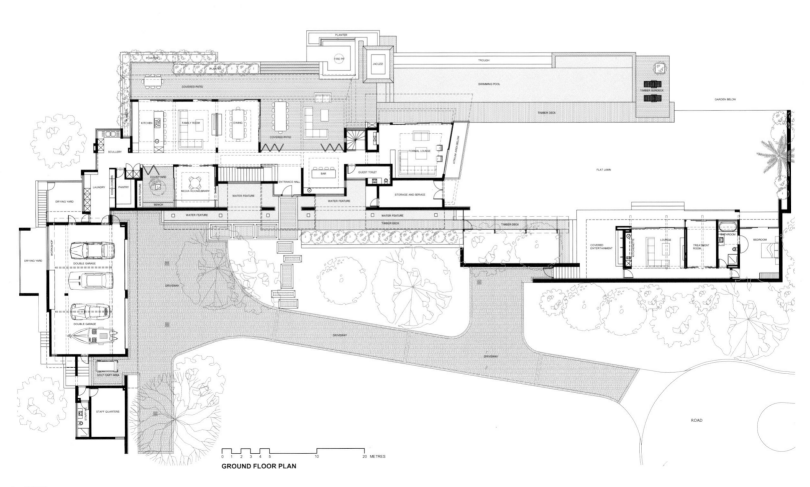

GROUND FLOOR PLAN

现代的家庭住宅合欢屋坐落于 Simbithi 牛态村的山坡上，占地 1 公顷。设计简约而精致既是客户的要求，也是合欢屋的特点。

在房屋所有的起居空间和卧室内都能观赏远景，斜坡上的茂密森林也在视线之内。

天然材料随处可见，屏风、露台和墙板都采用了木材；混凝土遮板和石墙与醒目的建筑外形，给人以视觉上的振奋感，而空间也不乏舒适感和温馨感。

房屋设计大量融合了水元素，25 m 长的浅水池设有玻璃隔板，隔开了水体和地下室中的电影间。清澈透明的大水池设在小径一侧，日夜倒映着房子，给人以宁静感。

建筑风格深受美国建筑师约翰·劳特纳的"Googie"未来主义建筑风格的影响。"Googie"一词的起源可以追溯到 1949 年，当时约翰·劳特纳设计了具有显著的建筑特点的西好莱坞咖啡店——"Googies"。

"Googie"未来主义建筑风格是现代建筑的一种形式，也是未来派建筑风格的一个分支。其风格受到 20 世纪 50 年代美国社会潮流和市场的影响，反过来也代表了那个强调未来风格设计、汽车飞机文化，被称作太空时代和原子时代的社会和市场。

未来主义建筑的特色是其动感的设计形式，表现为上斜式的屋顶、弯曲的几何形状，以及大胆地应用玻璃、钢材和霓虹灯，这些精髓都蕴藏在合欢屋中。

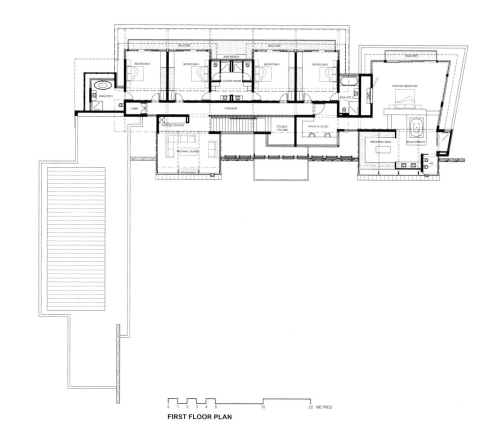

**FIRST FLOOR PLAN**

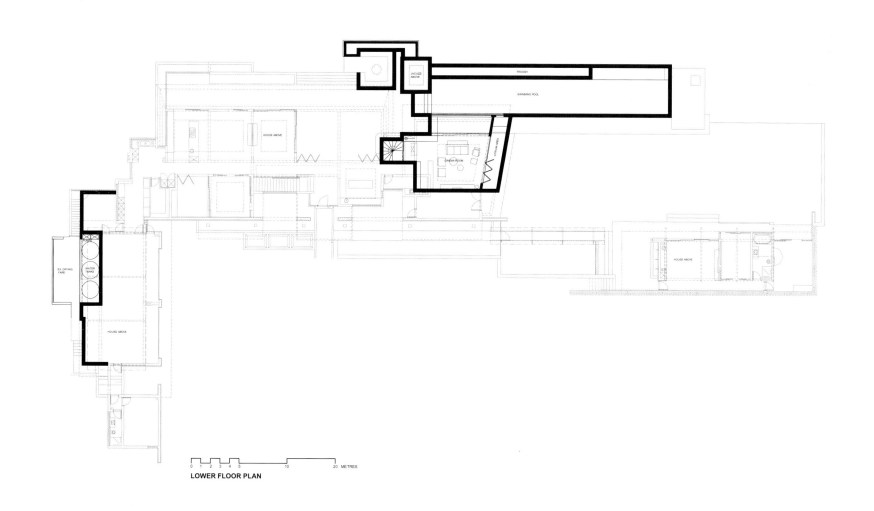

**LOWER FLOOR PLAN**

# Daniel's Lane Overview

长道盛景别墅

**Architects**
Blaze Makoid Architecture

"Our work focuses on creating total built environments that are a modern reflection of our clients while striving for a timeless product that remains fresh, exciting and inspiring."

The residence Blaze Makoid Architecture created for a father with three children in Sagaponack on the Eastern Shore of Long Island was inspired by the iconic architect Norman Jaffe's Perlbinder House(1970) and Tod Williams' Tarlo House (1979) but with his and his firm's signature of designing residences that have a quiet elegance that are uniquely suited to each client. As in all of Makoid's work, there is a cohesiveness that unites the architecture with its interiors and the site. The lines between indoors and out blur as they become the greater part of the whole.

The client put his trust in Makoid's ability to find the property and design a residence devoid of anything extraneous. His only mandate — he wanted a house that wasn't "busy".

Sited on a narrow, 4, 047 m², oceanfront lot, the design of this house was one of the first projects in the Village of Sagaponack to be affected by the 2010 revision to FEMA flood elevations, requiring a first floor elevation of approximately 5.2 m above sea level with a maximum height allowance of 12.2 m. All construction was required to be located landward of the Coastal Erosion Hazard Line. The location within a high velocity (VE) wind zone added to the planning and structural challenges.

Makoid wanted the structure to appear simple and clean upon arrival. The 2-storey travertine entry facade is highlighted with a single opening accentuated by a cantilevered afromosia stair landing that hovers off the ground. A "cut and fold" in the wall plane bends to allow for one large glass opening, from which an over scaled wood aperture containing the main stair landing cantilevers. A layer of service spaces run parallel to the wall plane creating a threshold prior to reaching the horizontal expanse of the open plan living room, dining area and kitchen that stretches along the ocean side of the house. 4.6 m wide floor to ceiling glass sliding panels maximize the ocean view and create easy access to the patio and pool beyond.

The second floor is imagined as a travertine and glass "drawer" floating above the glass floor below. 3 identical children's bedrooms run from west to east, setting a rhythm that is punctuated by a master bedroom with balcony. It is clad in the same afromosia wood as the stair landing. The quiet elegance and clean lines of the house are accentuated by the materials that also include poured-in-place concrete floors, Calcutta marble cladding and afromosia millwork.

"我们的工作重心是打造全新的环境，既拥有现代风格，又永不过时，永远维持新鲜、令人振奋、鼓舞人心的特质。"

该建筑是由 Blaze Makoid Architecture 工作室打造的。客户住在长岛东海岸的萨加波纳克小镇，是 3 位孩子的父亲。该别墅的设计灵感来源于诺曼·贾菲的 Perlbinder House（1970）和托德·威廉的 Tarlo House（1979）。当然，别墅也具有设计机构本身的特色，其幽静雅致的特点更是完全符合客户的要求。Makoid 打造的每个作品都具有室内空间与周围环境相融合的特点，通过模糊室内与室外空间的界限，使室外空间在整体空间中占据更大的比例。

客户相信 Makoid 有能力为自己找到与世隔绝的场地，并在场地上建造房屋。因此，他唯一的要求便是拥有一栋幽静的房屋。

建筑坐落在狭长的滨海区域，占地面积 4047 $m^2$。该项目是自萨加波纳克村受到 2010 年"菲玛"洪水的影响以来，首批建设项目之一。建筑首层按照要求，应当高出海平面约 5.2 m，而且高度限制在 12.2 m。整栋建筑都要求建在海岸侵蚀防护线内，并且该地的风速很快，增加了规划上以及结构上的挑战性。

Makoid 希望建筑结构给人的第一印象是干净利落。正面的墙壁用 2 层洞石装饰着，唯一的开口更是衬托了墙壁。悬臂式的红豆木楼梯在空间盘旋上升。壁板通过切割和组合，形成一个大的玻璃窗，悬臂式的楼梯也设在大木框架中。一排服务空间沿墙壁平行延伸，先是形成一道门槛，再与开放式的起居空间、餐厅和厨房相连，使这些面向海洋的空间在水平上显得更加宽敞。4.6 m 宽的滑动玻璃幕墙设置巧妙，使空间的海景达到了最大化，同时也可通往露台和水池。

第二层被构想成飘浮于玻璃地面上的由洞石和玻璃组成的"抽屉"。3 间相同的儿童卧室自西向东有节奏地展开，又被设有阳台的主卧室打断。此处的楼梯同样被红豆木包裹着。

建筑的选材，如现场浇筑的混凝土地板、加尔各答大理石以及抛光红豆木，凸显了房屋的平静优雅和清晰的线条。

/ 283

# Balcony House

阳台屋

**Architects**  
A-cero

**Location**  
Madrid, Spain

**Area**  
952 m²

The architecture studio A-cero, led by Joaquin Torres and Rafael Llamazares, presents one of its latest projects, Balcony House in Madrid.

This single family house has an area of 947 m² arranged on 3 levels and located on an individual plot of 2100 m² with a moderate slope and a prime location with natural surroundings, just 14 km from the center of the capital.

You access the house by the main door that stands out for its high altitude, settled in black glass and black gloss steel. On the ground floor (entrance floor), you enter through a hall with double height which separates the house in two main parts. On this floor, on the left wing, is located the most public and service area. The kitchen with pantry, living room area and stairs which connect all floors. On the right wing, you can find the master bedroom with dressing room and bathroom, and the guest bedroom.

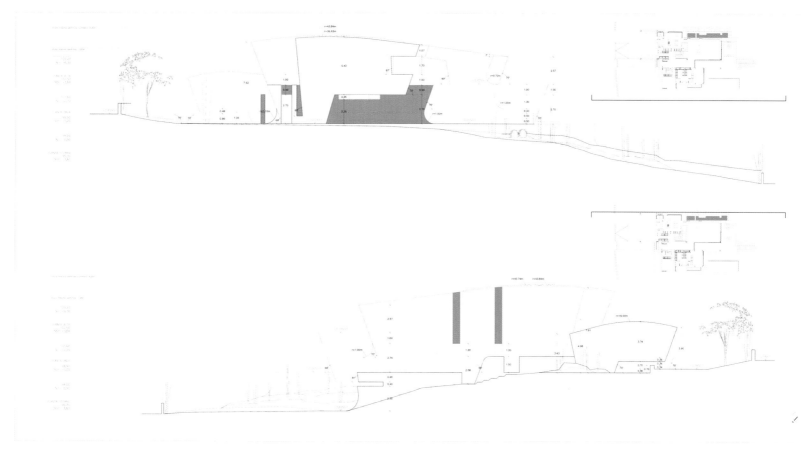

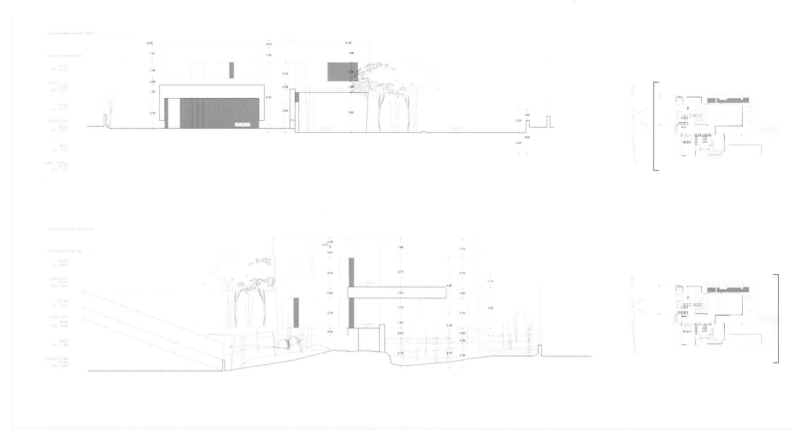

The living room features a great height that gives a sense of spaciousness. A central wood block runs through the ceiling to the wall and becomes television container with hinged panels in the same finish that ends in a fireplace.

This element also acts as a separator item for the dining room which has 2 tables for 8 people which can be combined into a table for 12. The whole room turns towards to the porch and pool area with large glass windows in black with openings, that allows a perfect interior – exterior communication.

The bedroom involves a practical and comfort philosophy and has the particularity that the windows open directly to the outside pool. They have the same views to the extensive wooded area that sets up the environment of the plot.

At the back you will find 2 generous closets open to a bathroom that stands out for its breadth, quality of finishes and elegant elements like the rest of the house.

Upstairs, the same double height foyer extends with a glass walkway that ends up in a bookstore. On one side, there are a study area, office, a cinema room and the two children bedroom suites.

The basement is reserved for installation rooms, utility rooms and iron, wine cellar and a large gym with a play area which

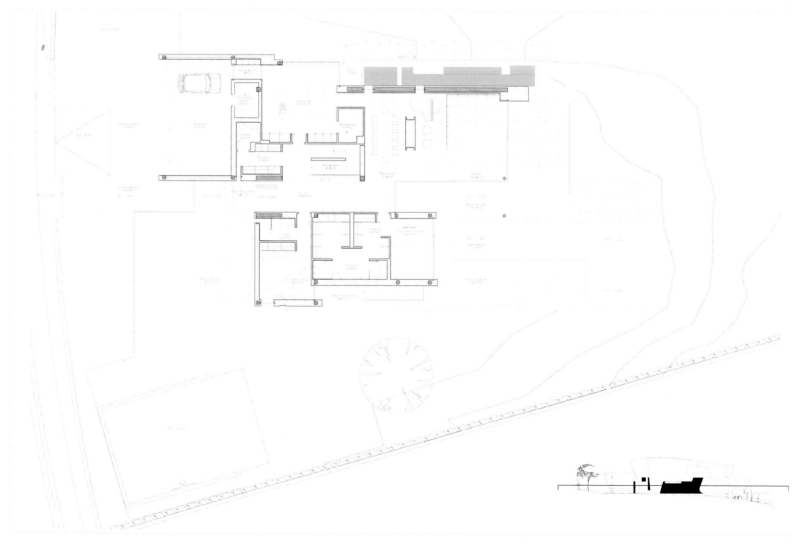

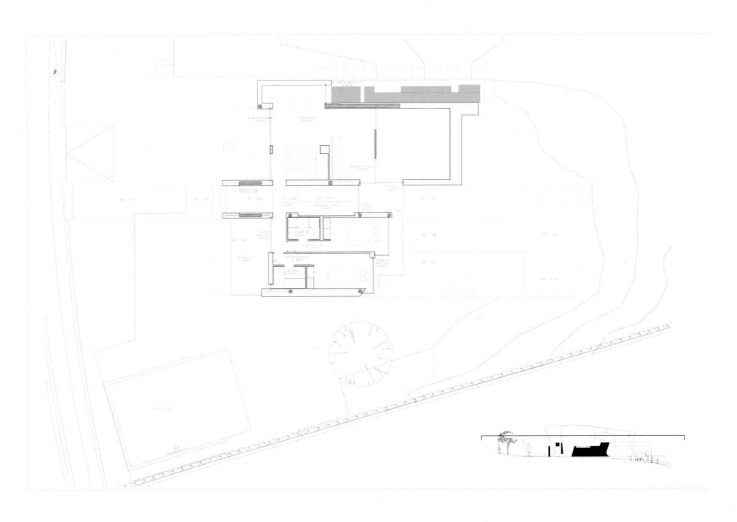
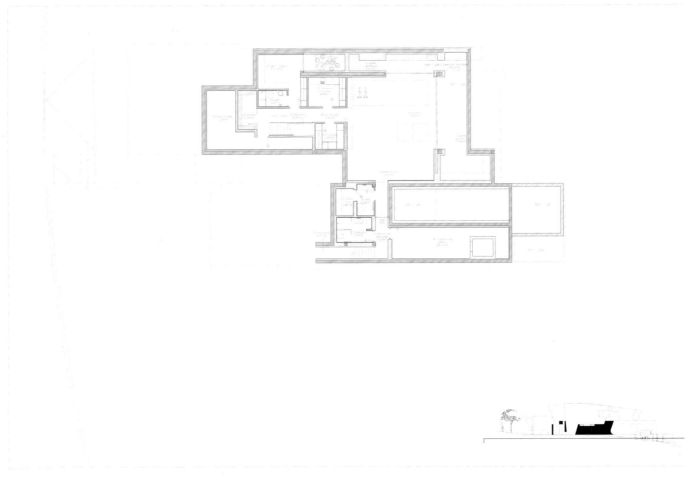

connects directly to a porch with a sliding windows also connecting this area with the garden. This porch is developed as a dining area, barbecue and outdoor kitchen, visually connecting the inside of the pool with the top floor.

The exterior of the house is designed with a curved sculptural image to add an organic value integrated with the environment, clad in white limestone and black glass. It represents an architecture rich in volumes, flat cuts and other aesthetic solutions which give meaning to the internal layouts and to the abundant light that floods into the house.

阳台屋位于马德里，是 A-cero 建筑工作室最新设计项目之一。该工作室是由 Joaquin Torres 和 Rafael Llamazares 两人负责的。

该家庭住宅面积为 947 m²，共 3 层，位于 2 100 m² 的场地内。场地坡度适中、地理环境优越，距离市中心只有 14 km。

高大的正门以黑玻璃和黑钢为主材。踏进大门，便是两层楼高的主大厅。空间分为 2 个区域，该层左翼主要是公共服务区，包括带餐具室的厨房、客厅和连接各楼层的楼梯。右翼则是配有更衣室的主卧室、浴室和客房。

客厅设在高层，给人以宽敞的感觉。木板从天花板延伸到墙面，再形成电视柜，进而延伸到壁炉。当然，电视柜上的铰接板与木板的外观是相同的。

餐厅的隔板也采用了木元素，餐厅内摆放了两张 8 人桌，餐桌还可以拼合成 12 人桌。整个空间朝向走廊和水池，黑色大玻璃窗上设有开口，使室内与室外能够更好地连通。

卧室的设计遵循了实用与舒适的理念。特别是窗户都直接朝向水池，使窗外广阔的林地景色尽收眼底，同时也营造了卧室的氛围。

房屋的后部设有两个大型衣柜，与浴室相对，就像房屋的其他部分一样，其宽度、质地和优雅元素使得它十分显眼。

同样，二楼设有两层高的客厅，客厅中的玻璃走廊延伸到了书房。一侧是学习办公区、电影室以及儿童双人套房。

地下室包括储藏室、杂物间、酒窖以及大型健身房。健身房的游乐区与走廊相连，推开推拉窗，走廊又与花园相通。走廊也可用作就餐区、烧烤和室外厨房，并把水池和顶层连接起来。

住宅的外观就像是雕塑，增加了有机感，也与自然环境相吻合。建筑外表则被白色大理石和黑色玻璃装饰着。此设计展示了一栋空间宽敞的建筑，同时还展现了平切及其他美学手法，使得室内布局更合理，室内光线更充足。

# Private Residence

隐蔽的豪宅

**Architects**
Robert Kolenik

A recent example of Eco Chic can be seen in the monumental villa in the south of Holland. The serene atmosphere of materials like wood and leather give a peaceful feeling of coming home. This big living is dominated by a stairway like we know from castles. As a soft tapestry in wood it unfolds itself into the room. In the kitchen area a splendid bar in rock crystal becomes a splendid eye catcher as it is lighted. In the living itself Kolenik designed a special fire that seems to float and which divides the room in a very natural way. The theatrical feeling increases with the stairs, now as a central monument in the room, with bending walls that high above split up in two symmetrical parts. The wooden steps make a very good contrast with the dark tiled floor in leather look. That makes a natural link with the rich leather furniture that invites to sit down. The open fire from Dofine in a crusted finish seems to float. It breaks the space with a look through from a cosy second sofa.

/ 295

With a long experience in design of professional kitchens Robert Kolenik knew exactly how the cooking area could be practical and beautiful, with most garments behind wooden doors in brush painted nut. There the beautiful lighted bar in smoked pebbles offers the intimacy of a pub without an association with cooking. If the cook is ready for it he or she finds everything in this custom made area. The newest appliances of Miele, an icecrusher, a wine storage and all the luxury to feel guest in your own house. In the cellar this estate did get a second bar where visitors can get a drink in privacy.

Taking the beautiful stairs brings us to the master bedroom and bathroom with stunning make up table. In a former cupboard het integrated the toilet in eucalyptus wood. The door handles in leather underline the soft feeling of the room and of the whole surrounding. Like in nature itself the combination of serene and special effects of lighting makes the owners every day happy as here they never have a dull moment.

该住宅坐落在荷兰南部，作为一栋不朽的别墅，它是时尚和生态的典范。建筑采用了朴素、沉静的材料，如木材和皮革，给人以一种宁静的家的感觉。房屋中最吸引人的是楼梯，与城堡中的楼道气势相当。柔软的木挂毯也在空间中尽显风采。厨房空间中设有壮丽的水晶吧台，被照亮时十分炫目。Kolenik 亲自给起居空间设计了壁炉，该壁炉十分特别，就像悬浮在空中，自然地将房间分隔开来。楼梯给空间增添了戏剧般的效果，它是房屋的精髓所在，将楼上弯曲的墙壁分隔成了两个对称的部分。楼梯上的木踏板与恰似皮革的深色地板形成巧妙的对比。古色古香的户外 Dofine 壁炉似乎在浮动，坐在第二个舒适的沙发上往外看，空间则更加开放。

Robert Kolenik 在专业的厨房设备设计上积累了丰富的经验，深知烹饪区域应当既实用，又美观。因此，橱柜都采用了木门，并用涂刷了的螺母来固定。灯光曼妙的独立吧台上镶嵌着烟灰鹅卵石，增添了吧间的亲切感，不会让人联想到烹饪。然而，当需要烹饪时，这个专门打造的空间能提供烹饪所需的一切。最新的美诺家电，比如刨冰机、贮酒设备等，给客人一种家的感觉。在酒窖中设有第二个吧间，客人可以独自享受美酒。

穿过美观的楼梯，便来到了主卧室和浴室，其间摆放的桌子也很有魅力。桉木把前柜与卫生间连接起来。裹着皮革的门把手突出了空间及整个周围环境的柔和感。就像自然界本身一样，融合了宁静和特殊灯光效果的空间，能让在此生活的人们充满快乐，永不枯燥。

# Russian Hill

## 俄罗斯山别墅

**Architects**  
jmA

**Principal**  
John Maniscalco

**Project Architect**  
Kelton Dissel

**Project Team**  
John Maniscalco, Kelton Dissel, Mick Khavari

**Location**  
San Francisco, CA

**Area**  
542 m²

**Photography**  
Paul Dyer

This certified LEED Platinum new 4-storey home establishes an understated but dignified urban presence on an atypically wide San Francisco site. A transitional 2-storey glass-walled entry hall draws users to an airy and open living level. An increasingly light stair element transitions from floor to floor ultimately arriving at a roof deck enjoying panoramic views.

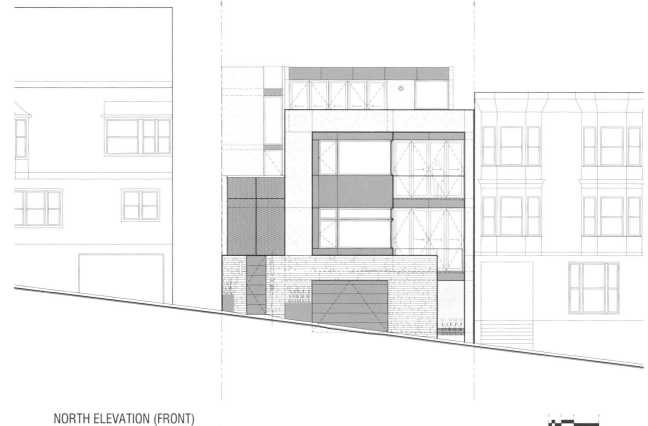

NORTH ELEVATION (FRONT)

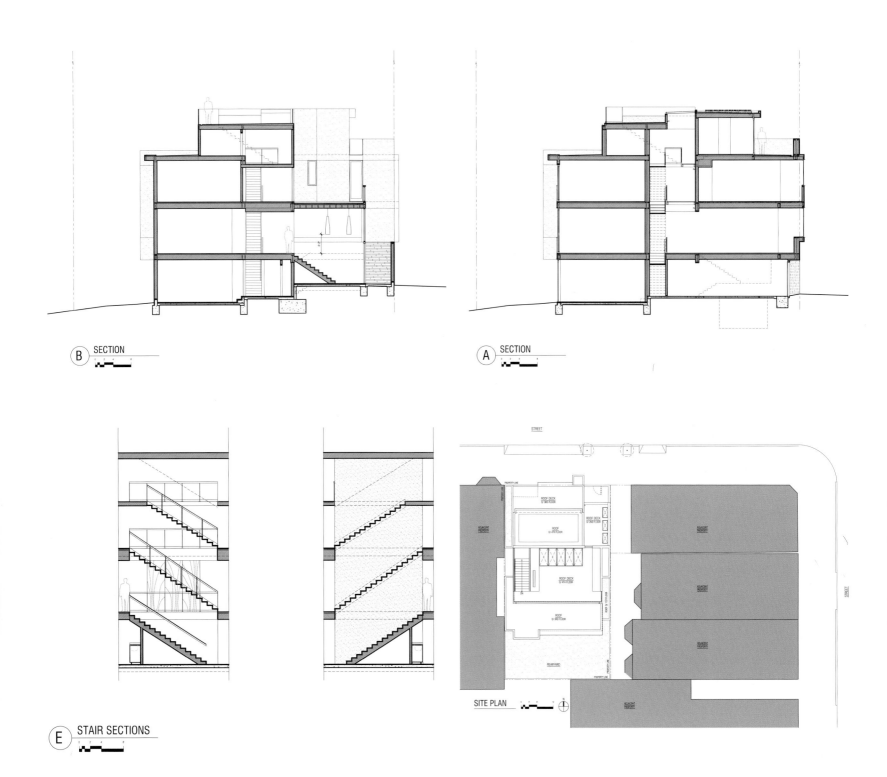

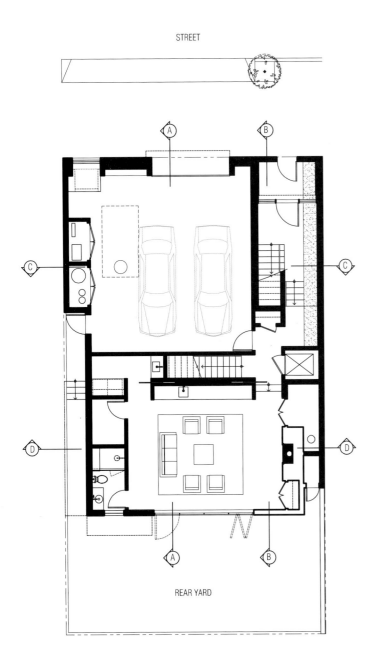

FIRST FLOOR PLAN

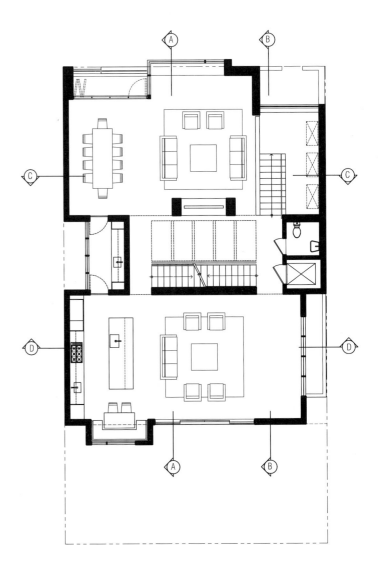

SECOND FLOOR PLAN

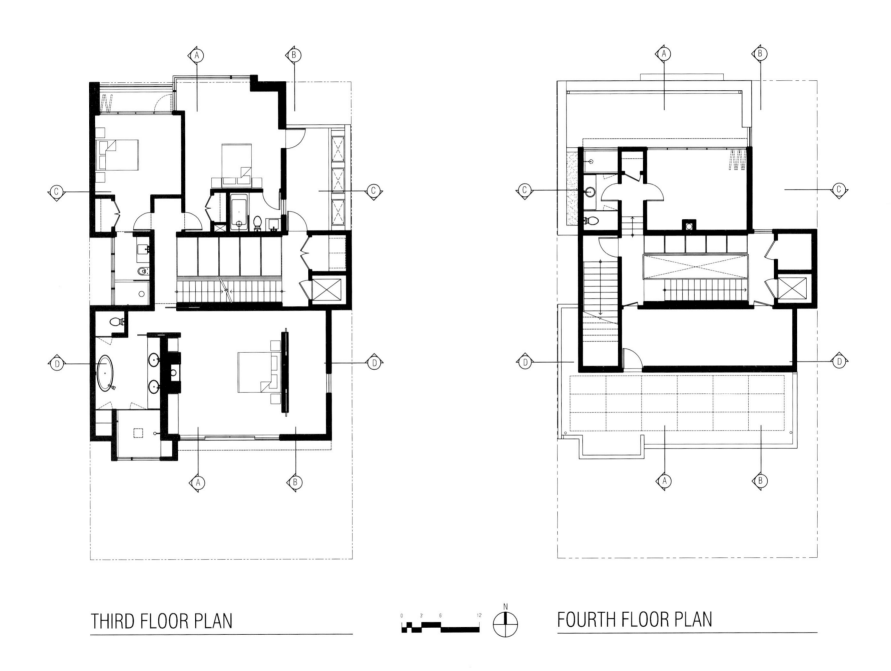

THIRD FLOOR PLAN

FOURTH FLOOR PLAN

该 4 层楼高的 LEED 铂金奖绿色建筑，赋予了异常宽阔的旧金山场地一种朴素而庄严的都市容貌。两倍高玻璃墙装饰着门厅，穿过门厅便来到了通风、开放的居住层。随着楼层的增高，楼梯间也越来越明亮，最后通向可以鸟瞰全景的屋顶露台。

# Casa Sorteo Tec 191

## 191号现代别墅

**Designer**
Arq. Bernardo Hinojosa

**Location**
Monterrey, N.L., México

**Area**
750 m²

**Photography**
Francisco Lubbert

The design reflects strong heavy materials which perfectly blend with the landscape that surrounds the house.

Steel and stone make an excellent material for décor and design of the property.

The pool and terraces make a relaxing environment and adding palm trees to the mix creates a resort like atmosphere.

The mezzanine library is a feature Arq. Bernardo Hinojosa likes to encourage, is one of many signature styles he has.

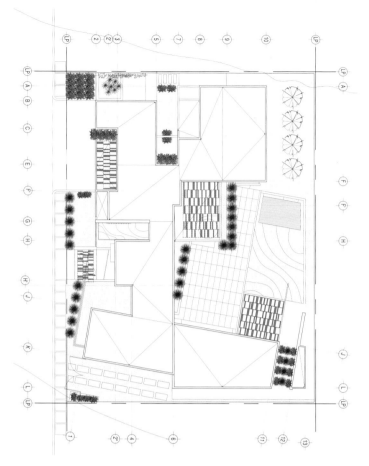

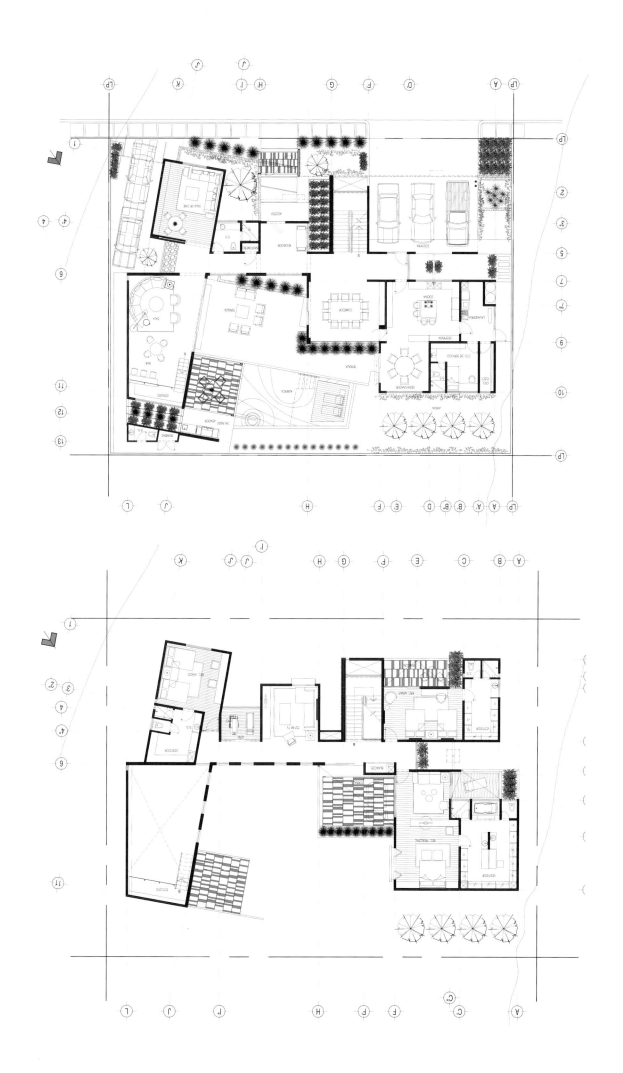

本项目采用了牢固的重型材料，实现了房子与周围景观的完美融合。

钢筋和石头是装修与设计别墅的绝佳材料。

住宅内设有水池和露台，使得别墅的氛围十分轻松，周围的棕榈则给空间一种度假村的氛围。

夹层图书馆是 Arq. Bernardo Hinojosa 推荐的特色，这是他的招牌风格之一。

/ 315

# Villa Escarpa

艾斯卡帕别墅

**Architects**  
Mario Martins

**Project Team**  
Sónia Fialho, Rita Rocha, Rui Duarte, Rui Saavedra, Sara Glória, Sónia Santos, Fernanda Pereira, Ana Filipa Santos, José Cabrita, António Caçapo

**Technical projects**  
Nuno Grave, Engenharia, Lda

**Client**  
anonymous

**Location**  
Luz Algarve, Portugal

**Photography**  
Fernando Guerra [FG+SG]

Villa Escarpa is located near the village of Praia da Luz, in the district of Lagos, Algarve, in the south of Portugal.

A condition of the planning permission was that the new house be constructed in the space occupied by a previous building. This had little architectural or technical merit, but was located in an exceptional position on an escarpment overlooking the Algarve coastline and village of Praia da Luz.

The footprint was therefore predetermined: on a very steep slope, and exposed to the prevailing winds. Paradoxically, it is these constraints and difficulties that underpin the conceptional basis of the project.

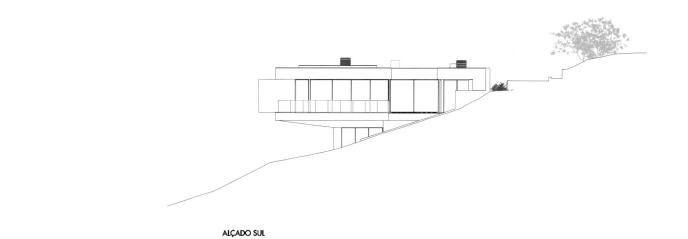

ALÇADO SUL

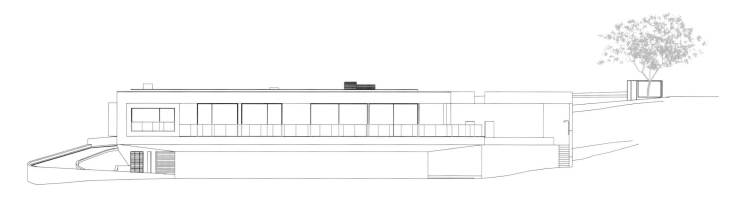

ALÇADO POENTE

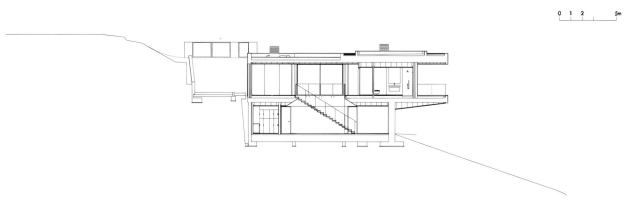

CORTE 1

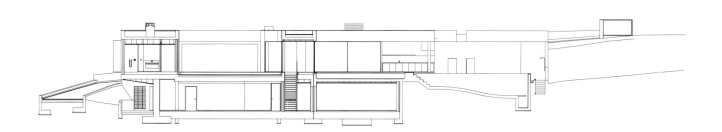

CORTE 2

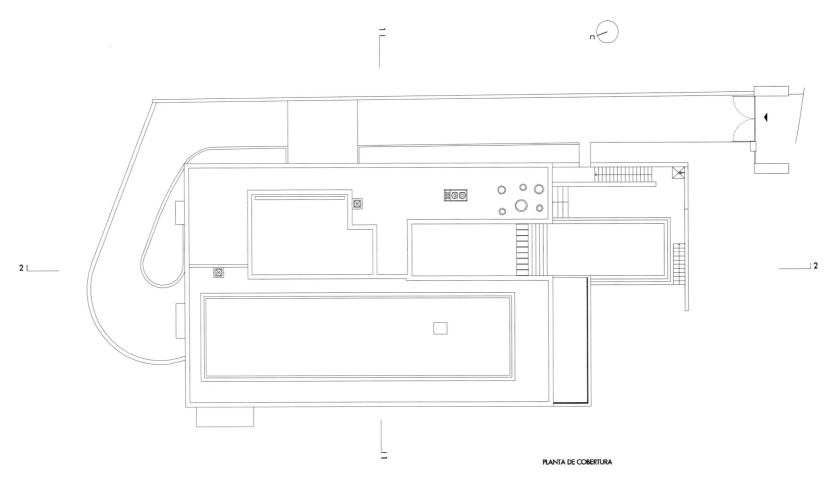

PLANTA DE COBERTURA

0 1 2 5m

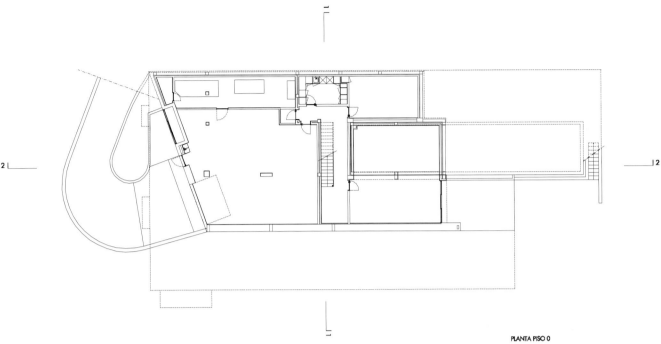

PLANTA PISO 0

In an architectural language, pure and contemporary, we created sheltered terraces and courtyards for outside living. These are cut from the horizontal volume which is white and highly transparent. This volume gently sits upon an exposed concrete support giving the appearance of the house floating above the landscape. The touch on the environment, which we want to preserve, is minimized and resolves the difficult balance of the building with its physical support. This ensures a desirable visual lightness.

The house merges with a long water surface which dissects the wide living and kitchen spaces. These spaces are complemented by terraces, open to the sun and impressive views. This is the social area of the house, open and fluid.

4 bedrooms are located in a private area with access from a corridor that runs alongside a central courtyard. In this private courtyard the natural light is filtered, creating an intimate and desirable space.

The lower area provides garaging and technical support.

The roof terrace accentuates the visual lightness of the floating building in its environment.

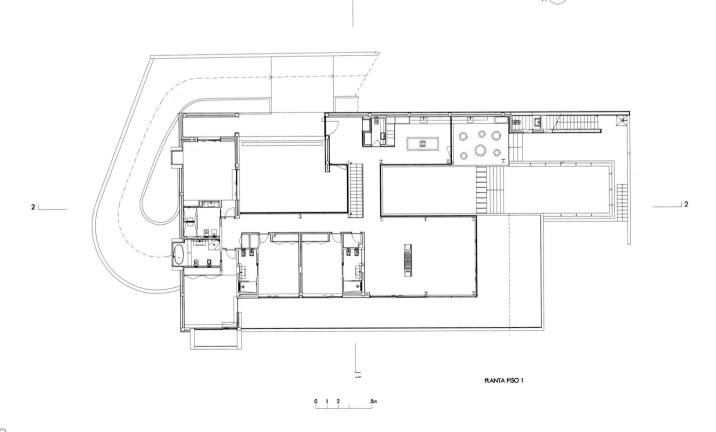

PLANTA PISO 1

艾斯卡帕别墅位于葡萄牙南部的阿尔加维拉各斯区，临近普拉亚-达卢什村。

建筑规划的前提是：新房子必须建在原建筑所在的场地上。这样一来，建筑上和技术上都不占优势。然而，房子坐落在悬崖的特殊位置，从这里可以鸟瞰阿尔加维海岸线和普拉亚-达卢什村。

为此，建筑的位置固定在十分陡峭的斜坡上，曝露于盛行风中。自相矛盾的是，正是这种局限和难题奠定了项目的理念基础。

以一种纯粹、当代的建筑语言，来为室外生活打造遮阳露台和庭院。这些空间都是从横向空间中分割出来的，采用了白色调，而且都是开放式空间。建筑坐落在暴露的水泥基座上，就像是悬浮在周围一切之上。在本项目中，建筑对环境的影响达到了最小化，并用建筑本身的支撑来解决了建筑的平衡问题，而且空间的亮度也达到了理想的状态。

长长的水池把宽敞的厨房与起居空间分隔开来，却与房屋连为一体。这些空间都敞向露台，阳光充足，风景壮丽。这个开放而流畅的空间就是社交空间。

私人区域中设有4间卧室，从中央庭院延伸至此的走廊可以通往各个房间。照射到私人庭院中的自然光经过过滤，使空间更加亲切而合意。

楼下区域提供了车库和技术支持。

楼顶露台突出了建筑，使建筑浮于周围环境之上，十分引人注目。

# Oceanique Villas

海洋别墅

**Architects**
MM++ architects / MIMYA .co

**Project Architect**
My An PHAM THI

**Location**
Mui Ne, Phan Thiet, Vietnam

**Building Area**
1,014 m²

**Photographs**
Hiroyuki OKI

This project is a small real estate development located in Mui Ne, a seaside holiday destination in Vietnam's south east coast.

The site has a trapezoidal shape 110 m deep between the ocean and the road and 42 m along the beach. The sea front view is amazing and unusual for a private residential project in this area, usually occupied exclusively by resorts and hotels. The idea was to maximize this potential with semi-detached sea front villas and keep a large part of the land as a "buffer" landscaped area to prevent noise from the road.

The construction is composed of 3 units: Two 3 bedrooms villas, and one 4 bedrooms villa, each with a private 10 m x 3 m swimming pool. The construction was raised to 1.8 m from the beach level in order to keep a good privacy from the public beach, maximize the sea view and prevent from site erosion.

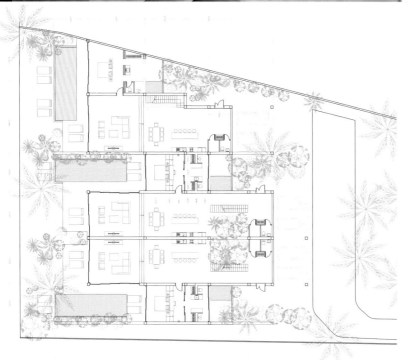
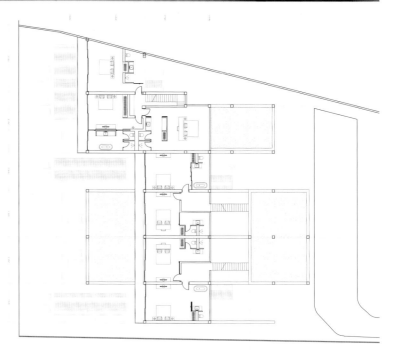

The interior layout offers a generous sea view for each space:

On the ground floor the kitchen, dining and living rooms are merged in one single space mono-oriented to the beach front while the back has been designed with a patio open to the sky to bring sunlight, natural ventilation and a nice sequence for entrance and staircase with the pond and the interior tropical garden.

Master bedrooms are open to the infinity swimming pool and by extension to the sea. In the back, each bathroom has a private garden with Jacuzzi, bringing light, ventilation and strong outdoor/indoor connection.

On the first floor, all bedrooms have a large ocean view with floor to ceiling windows recessed in the concrete structure providing enough shadows. On the front, the flat roof becomes a huge timber decking terrace, extending possibilities to contemplate the ocean.

In the back, are located the bathrooms, naturally ventilated through timber louvers.

Elevations are simple, following the layout strategy. The beach front elevation is widely open with large aluminum frame windows while the back elevation has no windows but timber louvers only. Therefore, in addition to the open air patio on the ground floor, it has been possible to limit the air conditioning to the bedrooms only.

This amazing location, so closed to the sea, is a dreaming place for living. This project tries to not miss this opportunity to offer to the guests a total and unique "seaside" experience.

In line with our practice, the materials are basic and the design is minimalist, focusing on the quality of each space with no sophistication.

该项目是开发位于越南东南海岸美奈海边度假胜地的一栋小房产。

场地呈梯形，介于海洋和道路之间，距离海洋 110 m，面向海洋一侧为 42 m。迷人的海洋风景对于这片区域的私人住宅来讲是极为少见的，通常只有度假村或酒店才会有。设计目的在于通过半独立式海滨别墅来最大化利用这种优势，同时保留大面积的"缓冲"景观区，隔离街道上的噪音。

建筑结构分为 3 部分：两栋三房别墅、一栋四房别墅以及一个 10 m x 3 m 私人泳池。建筑底座高出海滩 1.8 m，不会被公共海滩看到，从而保护了住宅的隐私，同时也扩大了海景，阻止了场地的侵蚀。

内部布置恰到好处，每个空间都有丰富的海景：

一层是厨房、餐厅和客厅，这些空间融为一体，面向海洋，背靠露台，使得空间内光线充足、自然通风，入口和楼梯设置合理有序，水池和室内热带花园更是迷人。

主卧室敞向无边泳池，并延伸至海洋。后方的每间浴室都带有私人花园和按摩浴缸，而且十分明亮，自然通风，实现了室内与室外空间的对接。

二层的所有卧室都设有玻璃幕墙，能欣赏广阔的海景，玻璃墙嵌入到混凝土结构中，形成了足够的隐密空间。房屋前端的屋顶被建成宽阔的露台，可供观望海景。

屋后部的浴室装有百叶窗，自然通风。

房屋正面设计十分简单，遵循了一定的布局策略。面向海洋一侧是铝框大窗，而另一侧不设窗户，只有木百叶窗。因此，除了一层的露台以外，只有卧室需要用空调来调节室温。

房屋是如此地靠近大海，令人难以置信，这似乎是梦中才有的生活空间。项目最大程度地给人提供一种完全独特的海边体验。

设计具有极简主义风格，不玩弄复杂手法，注重每个空间的质量。与设计同样重要的还有材料基础。

# CONTRIBUTORS 设计师名录

### A-cero

A-cero joaquín torres & rafael llamazares architects is an architecture firm founded in 1996, vowed to the integral development of architectural, interior design and urban planning projects.

The studio has a team of more than 100 specialized professionals, led by architect Joaquín Torres Vérez, working in two main headquarters in Spain, located in Madrid and A Coruña, and a foreign projects office in Dubai.

The practice has evolved in parallel to its clients' demands, building a large portfolio that comprises residential estates, resorts, office buildings, mixed-use high-rises, corporate and interior design. We have the capacity to carry out an architecture project in all its stages, since its conception until the construction management.

Our working method is based on a detailed analysis of the client's needs and the project's program. The result is examined in all its dimensions in order to find every potential problem, all the possible solutions are studied and all the material possibilities for the new building analyzed during the design stage. The concept ideas are kept from the beginning to the end, as a conducting thread for the design process, in order to achieve the best possible results.

Passionate and motivated, A-cero designers develop innovative and visionary designs for our clients. Driven by our core values of individuality, respect, passion and integrity, our commitment to design excellence is integral to everything we do. At A-cero we want to experience the pride and excitement of creation and it is our commitment to be the very best that drives us forward.

### Arch. Bernardo Hinojosa / Associates

Based in Monterrey, México, this architectural practice has a diversified practice, with over 2,500,000 m² designed in Master Planning, University Buildings, Private Houses, Industrial, Religious and Commercial Architecture. It is consider on of the top 5 most important architectural firms in Northern México.

More than 40 international, national and local prizes have been obtain for his distinct projects as well as publications of the web, architectural magazines and several books. His work has being exhibited in Netherlands, Belgium, Dallas, New York and Athens.

Bernardo Hinojosa's architecture seeks "atemporality", which states all projects should look contemporary even with the passing of time, and it is characterize by its functional and rationalist bases.

We embrace any challenge to overcome it, with an excellent piece of architecture as a result.

### Belzberg Architects

Belzberg Architects is about architectural innovation.

Design is a collaborative process, a dialogue, between client and architect.

Form is a merger of conceptual investigation with production methodology.

Practice is flexible, nimble, evolving, and unconventional.

### Blaze Makoid Architecture

Blaze Makoid has been practicing architecture and design since graduating from Rhode Island School of Design in 1985. His company Blaze Makoid Architecture, located in Bridgehampton, NY, was established in 2001 and since its inception has created sophisticated, luxury residential architecture in the most sought after locations.

The firm's work has been recognized in The New York Times, Architect magazine, Hamptons Cottage & Gardens, Beach, Ocean Home, and the Robb Report for designs that acknowledge the lifestyle and day-to-day experience their clients' desire on beautiful, yet demanding sites.

Blaze and the firm have received numerous national and international design awards, including The Long Island AIA Commendation Achievement in Residential Design; The AIA Peconic Honor Award; The Boston Society of Architects' First Citation; and the Philadelphia AIA honor Award for Excellence.

### Design studio Yuri Zimenko

Eponymous design studio, designs and design supervision of design projects: private houses, apartments, hotels, restaurants, cafes, offices, shops, as well as product design. In developing the project takes into account all the wishes of the customer.

Studio work has repeatedly won prizes at various competitions and many times were marked attention from the press.

### GREGWRIGHT architects

GREGWRIGHT architects (GWA) was founded in 1995 by its sole member at the time, Greg Wright. The decision to break ties from a successful partnership with my previous partner and form a new practice was formed out of a need/desire to explore a new and separate vision for the making of buildings and to discover a different way in the delivery of both as a service and product that adhered to the principles and values held as important to Greg Wright. These principles still underpin the ethos of the practice and include:

• The business of architecture should be inspired by the delivery of contemporary buildings founded on the values of integrity, honesty, and a commitment to offer positively to the built environment

• The belief in and commitment to ones ideas is the single most valuable currency we can invest in

• Work is driven by a desire to make buildings that reveal & revel in excellence of thought and execution – this soon became the credo of the newly formed GWA.

The practise continues to develop projects around these core values. A holistic approach is adopted for all projects and this often includes the design of interiors and bespoke one-off items for their clients. This approach has often resulted in projects being critically acclaimed and featured in various publications both locally and internationally as well as TV programmes such as our local Top Billing.

From humble beginnings of doing primarily residential work, the practise grown both in confidence and size taking on bigger challenges and completed the recent master plan for the Centenary City of Abuja, Nigeria together with key landmark buildings designs. Our work continues to expand into Africa with submissions to the Ivory Coast, work in Namibia and current new commissions in Luanda Angola.

In 2001 Greg Scott joined GWA and very quickly became a pivotal part of the success of the practise. This was acknowledged in his appointment as a director and shareholder in GWA in 2003, and has been instrumental in the growth of GWA and the extensive portfolio of work completed to date.

## John Maniscalco

John Maniscalco, AIA, is the founding principal of JMA. He brings 25 years of professional experience as a Project Designer for a wide range of projects including master-planning, commercial office buildings, academic facilities, civic centers, transportation complexes, and wineries, as well as multi and single-family custom homes. He is a graduate of Cornell University and has worked as a Project Designer in the offices of Gensler, Chong Partners, ROMA Design Group, and Ellerbe Becket. Since starting JMA in 2000, he oversees all phases of design and construction on JMA projects from conception to completion. His work has received numerous design awards and honors, and has been featured in national and international publications.

## Keith Baker Design

KB Design (Keith Baker Design Inc.) was established by Canadian designer Keith Baker in Victoria, B.C. in 1989 as a residential design practice specializing in the design of unique custom homes and well integrated renovations and additions. KB Design is a member of the Canadian Home Builders Association. Keith's work has been recognized over the years with an impressive 22Gold awards and 47 Silver awards on the local, provincial and national levels for excellence in categories ranging from Project of the Year, Best Custom Home in Canada, Best Custom Home, Best Master Suite, Best Interior, Best Bathroom, Best Renovation, Best Kitchen and the coveted People's Choice Award. His work has been widely published in local, national and international books and periodicals and featured on hundreds of websites profiling Contemporary West Coast and Modern residential design.

The company is a three person team consisting of two designers and one building technologist and enjoys the close client relationship afforded by a small high quality residential design practice.

Keith accepts design commissions in his home of Victoria, B.C. as well as across Canada, the United States and internationally.

Keith and his team, using state of the art 3D design technology coupled with his thirty plus years of professional design experience, are able to create unique and varied custom home design solutions.

Keith believes strongly in the need for green and energy efficient home design as well as the concept of biophilic design   the idea of the use of natural building materials and a profound connection with the natural world creating beautiful healthful living environments.

## Kolenik Eco Chic Design

Robert Kolenik (b.1981) is one of the representatives of the new generation of Dutch Designers who are renowned worldwide for their down to earth mentality, excellence and innovative approach to design. In 2005 Robert took over his father's business and subsequently Kolenik Eco Chic Design was founded in 2008.

With a dedicated team of experienced architects, the multidisciplinary studio is responsible for a wide range of projects. From bespoke designs for private villas and product designs through special fields of interest to the development of elegant and luxurious total concepts for the hospitality industry. More recently the studio earned a great reputation working on several hotels, restaurants, bar and club designs all with their signature international and timeless allure.

The distinct and recognisable Kolenik style appeals to a wide international audience. It is best described as minimalist and warm, carefully marrying functionality and aesthetics to create harmony and balance.

Kolenik hotel and villa designs exude an inviting quality of peace and tranquillity, while restaurants and other hospitality designs are more daring, aiming to fascinate and excite their guests.

The use of materials like luxurious natural stone and even living nature add interest to their spaces and are the hallmark of Robert's style.

Each and every Kolenik design is unique; from the made to measure furniture to the most minute, beautiful details and innovative solutions.

The name Eco Chic Design deliberately reflects the designer's passion for nature.

Together with his team he tirelessly searches for original, smart and intelligent but sustainable solutions to take their work to a higher level.

Robert Kolenik is also co-founder of Plastic Soup Foundation Junior, a charity that educates children at a very young age about the dangers of polluting our the seas and oceans with plastics and its consequent impact on marine life.

## Mário Martins Atelier

The company, mário martins - atelier de arquitectura, lda, based in Lagos, Algarve, Portugal, has been working in the field of architecture and urban planning for more than 20 years.

The company consists in a diverse team of permanent staff, a group of technical experts, who work together in the different areas and specializations of projects.

The company has been extremely busy over the past decades, which has resulted in a considerable volume and diversity of projects built: single family houses, collective housing, residential condominiums, tourist developments (hotels, aparthotels, tourist apartments, etc), restaurants/bars, various public facilities (sports, social, educational, recreational, etc), urban renewal and planning.

The projects have been recognised with prizes, nominations and publications over the years in Portuguese and international publications (magazines, books and online).

This presentation is based on houses (new and recovered) done during last 15 years is this area of the Algarve, Portugal following the concept: Respect for the site, environment, local culture and the people.

Reinterpretation of these aspects to archive a sustainable and contemporary result.

## McClean Design

Paul McClean trained as an architect in Ireland and founded McClean Design in 2000.

Over the last fifteen years, MCCLEAN DESIGN has grown into one of the leading contemporary residential design firms in the Los Angeles area committed to excellence in modern design. We are currently working on more than twenty large homes across much of Southern California with additional projects in San Francisco, Vancouver and New York.

Our projects reflect an interest in modern living and a desire to connect our clients to the beauty of the surrounding natural environment. We make extensive use of glazing systems to maximize views and provide a warm light filled contemporary space. We strive for simplicity and an openness to the surrounding landscape that erodes the division between indoor and outdoor spaces; homes with an emphasis on texture and natural materials.

We are committed to environmentally sustainable design practices and have extensive experience in both Orange and Los Angeles counties with a proven ability to navigate complicated approval processes such as the Laguna Beach and Beverly Hills Design Review Board as well as Coastal Commission. We keep an open mind on questions of style preferring to strike a balance between the best solution for the site, our clients' preferences, and what is potentially approvable for each particular site.

We work with both homeowners and developers. Our staff offers a full range of design skills ensuring that our projects are completed in a timely manner and to the highest standards.

We continue to strive for excellence in design and to push the boundaries of imagination in creating extraordinary spaces that we hope will provide enjoyment for many years to come.

## MCK Architects

MCK is a young team of multi-award winning architects based in Sydney, Australia.

They specialise in residential and commercial projects of high-quality finish and detail that are sensitive to context and brief. Their distinctive aesthetic is known for its classic proportion and geometric form.

We enjoy using unexpected materials that challenge and sometimes even surprise. We also believe that good architecture and respect for the environment go hand in hand.

## Metropole Architects

Metropole Architects was founded in 1997 by Nigel Tarboton. Tyrone Reardon joined the office in 2002, and became a partner in 2004. Our office is situated in Durban, Kwa-Zulu Natal, South Africa.

As a practice, Metropole Architects, are inspired by the energy of the city, as it unceasingly moves, radiates and evolves like a vast living organism. We aim to generate design that stakes out new territory, and explore ideas that are intuitive, inventive, exuberant and daring. Enthused by visionary architects like John Lautner and Santiago Calatrava, we aspire to create iconic and progressive architectural wonders that capture our collective imagination.

## MM++ architects

My An Pham Thi, Architect, Graduated from University of Hanoi. After more than 10 years of practice in different international offices she founded Mimya co. (MM++ architects) in 2010. Michael Charruault, Architect, Graduated from the French architecture school Paris-Belleville. Based in Vietnam since 2007, co-founder.

## Nico van der Meulen Architects

As an architectural practice that is well renowned throughout the African continent, NICO VAN DER MEULEN ARCHITECTS supplies creative solutions that are customized to suit each client's personal requirements. Through working closely with all its clients to ensure optimal satisfaction, the practice has accomplished astounding success in the design of upmarket residential homes. The company's innovative architectural vision is evident in its ability to continuously produce outstanding and artistic architectural designs that are personalized in accordance with the homeowners' lifestyle requirements.

## OPENSPACE DESIGN Co., Ltd.

OPENSPACE DESIGN is a Thailand based architecture & interior design company founded in 2005. Their multidisciplinary design team is composed of architects, interior designers, graphic designers, R&D specialists. Professionally, they work together in the synchronized manner enabling them to serve their clients completely.

There are 4 main types of design services provided as follows: Architectural Design, Interior Design, Landscape Design and Graphic Design.

OPENSPACE DESIGN has accumulated design proficiency and experience through various types of projects such as office, commercial building, shopping mall & retail, hotel-resort-spa, residence (housing/condominium), educational building, hospital, master planning & landscape, etc. Their services cover projects both in Thailand and abroad including government's and private's sectors.

OPENSPACE DESIGN provides full package of design services from the client's requirement analysis, project's feasibility study, design strategy to achieve the uniqueness of the project-undoubtedly, several different design alternatives are meticulously compared before creating the final one, working with our client closely until the project's completion. With their service-mind and true determination, they always make sure that each project meets high efficiency, client's satisfaction and the users' aesthetical experience.

## Pitsou Kedem Architects

Pitsou Kedem Architects was founded in 2002 and today employs nine architects (Pitsou Kedem, Irene Goldberg, Nurit Ben Yosef, Raz Melamed, Noa Groman, Ran Broides, Hila Sela, Tamar Berger, Emanuel Amsalem).

The office was established by Pitsou Kedem, a graduate of the Architectural Association in London and mentor of final projects at the Technion Haifa's Faculty of Architecture. In the past two years, the office has received five awards in the Israeli "Design Award" competition, and has been chosen Architect Office of the Year in the "Private Construction" category by Israeli Construction and Housing magazine.

The office designs private as well as commercial projects such as B&B Italia's Tel Aviv flagship store, a boutique hotel on the city's prominent Rothschild Boulevard and an events hall.

## Rolf Ockert Design

At Rolf Ockert Design we continuously aim to find individual and optimised solutions for any design task we can get our hands (and minds) on.

Having lived and worked in Europe, Asia and Australia and having travelled the world extensively Rolf draws on a wide range of visual, practical and design experience.
For us it is still fascinating how different architectural solutions for the same task can be. The solutions then are a reflection of the designers experience, taste, preference, etc. We call it the mosaic of life, every little bit of input forming a vital part of the whole piece.

Rolf Ockert Design, commenced in 2004, has in a relatively short period of time created a rich portfolio of work, ranging from high end residential design to commercial and retail projects, product and furniture design, master planning and much more. Projects to date have been located throughout Australia as well as overseas, most recently in New Zealand, Japan and Switzerland. Many projects have been published in national and international publications.

## SAOTA

SAOTA is a firm of architectural designers and technologists including in-house CGI and marketing teams, as well as a strong support staff. It is driven by the dynamic combination of Stefan Antoni, Philip Olmesdahl, Greg Truen and Phillippe Fouché who share a potent vision easily distinguished in their design. This, paired with both an innovative and dedicated approach to the execution of projects, has seen SAOTA become internationally sought-after, receiving numerous awards and commendations from some of the most respected institutions worldwide.

Capitalising on a unique understanding of an ever-evolving industry, SAOTA continues to build on past experiences and is well positioned to offer expert services to the corporate, institutional, commercial and residential marketplace.

With roots in South Africa, SAOTA now has an international footprint with projects on five continents.

## Stuart Narofsky

Stuart Narofsky, AIA, LEED AP is the principal of Narofsky Architecture, a multi-discipline design firm founded in 1983 that also offers design build and interior design services. His partner Jennifer Rusch oversees ways2design, the interior division.

Stuart is former President of the American Institute of Architects-Long Island Chapter. For ten years he was an Associate Professor of Architecture at New York Institute of Technology, overseeing a design studio and teaching furniture design. More recently as a visiting Professor of Architecture at Pratt Institute, where he taught an upper class Design Studio, specializing in modular construction. Stuart has over the past three years he has been visiting a professor in workshops at the faculties of Architectures at Universities in Argentina (La Plata) and Bolivia (Santa Cruz, Cochabamba, Sucre and La Paz), on the subjects of Sustainable Architecture and practicing architecture in New York.

His projects have received many awards: Architectural Digests best Home Competition; Queens Chamber of Commerce Award for his competition winning scheme for a Ronald McDonald house; numerous architecture awards from the AIA, including AIA Long Island Chapter, where his Patel residence garnished multiple awards in residential design, and best project of the year; the Society of American Registered Architects; and AIA New York State.

Recently, a single family house on Long Island obtained the 2012 Long Island AIA Chapter Archi Award for a sustainable residence.

## The Agency

The Agency is a full-service, luxury real estate brokerage representing clients worldwide in a broad spectrum of classes, including single-family residential, new development, resort and hospitality, residential leasing, luxury vacation rental and property management. The Agency was founded by Mauricio Umansky, who was recently recognized by The Wall Street Journal as the #1 top-producing real estate agent in Southern California and #7 in the U.S., and design and architectural specialists Billy Rose and Blair Chang, whose Rose + Chang team was named multiple times as one of the Top 100 U.S. real estate sales teams by The Wall Street Journal. Shunning the traditional brokerage model of cut-throat agents competing against each other, The Agency fosters a culture of partnership in which all clients and listings are represented in a cooperative environment by all its agents, thereby ensuring its clients and listings have the competitive edge. Leveraging the most emergent technologies and social media strategies, The Agency envisions itself as more than just a real estate brokerage; it is a lifestyle company committed to informing and connecting global communities. The Agency tailors global marketing solutions for buyers, sellers, developers and landlords

## Urbane Projects

Urbane Projects was founded in 2003 by Managing Director, Steve Gliosca.

Urbane Projects has established itself as a premium boutique builder in WA, specialising in designing and building luxury homes across Perth with a focus on providing every client with personalised, tailored attention.

Urbane Projects' Managing Director, Steve Gliosca, is both the designer and builder – a rare luxury in today's construction industry which ensures the homes coherence and success from conception to implementation.

The goal at Urbane is to provide every client with an enjoyable, stress-free building experience and it speaks volumes about their quality of service that almost all of their business generates from word of mouth.

The distinguishing feature of building with Urbane Projects is "Our People". The team at Urbane Projects offers exceptional service, complete transparency and our in-house solutions provides each client with a combined skillset of a designer, interior architect, estimator and builder all under the one roof. Of course, Urbane recognise that without clients there is no product and they are privileged to have worked with a vast array of people that now formthe Urbane family of clients. It is their dedication and trust in Urbane Projects that have enabled the design and construction process to be pushed to the limits – to strive to set new benchmarks that serve as an inspiration to others.

Urbane Projects offers a complete turnkey package. From design conception to completion we assist every client with individual detailing of all key selections for both the exterior and interior of the home. Our homes have distinctive curb appeal whilst ensuring a floor plan that is unique to each client and their individual lifestyle.

## Whipple Russell Architects

Marc Whipple, the founder of Whipple Russell Architects, is the son of an American Diplomat, Marc Whipple grew up across Europe, Asia and Africa, whose rich cultures helped to shape his eclectic approach. Following his education at  Eton College and London's prestigious Architectural Association School of Architecture, he became the protégé of internationally renowned architect George Vernon Russell. Russell, creator of show-stoppers like the Trocadero on Sunset Boulevard, the Flamingo in Las Vegas, as well as Samuel Goldwyn's home in Beverly Hills and the expansive University of California at Riverside campus, further broadened Marc's vision.

Twenty-five years ago, when Marc opened his own firm, he honored his late mentor by including his name in that of the practice. Since that time, Marc Whipple has demonstrated a range of scale and innovation that extends from intimate west coast life-style specific homes in the Hollywood Hills to the Sienna Hotel Spa Casino in Reno to a master plan for an island-spanning resort in the Caribbean. His firm, whipple russell architects, is noted for applying authentic materials, natural light and green technology to the marriage of elegant form and efficient function.

Whipple Russell architects, formerly known as The Russell Group Architects, has been featured in periodicals that include Metropolitan Home, Dwell Maga zine, Robb Report, the Los Angeles Business Journal, Home Beautiful, In Style Home, The Los Angeles Times and The New York Times.

[ARTPOWER]

## Acknowledgements
We would like to thank all the designers and companies who made significant contributions to the compilation of this book. Without them, this project would not have been possible. We would also like to thank many others whose names did not appear on the credits, but made specific input and support for the project from beginning to end.

## Future Editions
If you would like to contribute to the next edition of Artpower, please email us your details to: artpower@artpower.com.cn